Principles of Agroecology

Principles of Agroecology

Edited by
Milan Collins

Larsen & Keller
www.larsen-keller.com

Principles of Agroecology
Edited by Milan Collins
ISBN: 978-1-63549-019-0 (Hardback)

Larsen & Keller

Published by Larsen and Keller Education,
5 Penn Plaza,
19th Floor,
New York, NY 10001, USA

Cataloging-in-Publication Data

Principles of agroecology / edited by Milan Collins.
    p. cm.
Includes bibliographical references and index.
ISBN 978-1-63549-019-0
1. Agricultural ecology. 2. Agriculture--Environmental aspects. 3. Ecology. I. Collins, Milan.
S589.7 .P75 2017
577.55--dc23

The publisher's policy is to use permanent paper from mills that operate a sustainable forestry policy. Furthermore, the publisher ensures that the text paper and cover boards used have met acceptable environmental accreditation standards.

Printed and bound in the United States of America.

For more information regarding Larsen and Keller Education and its products, please visit the publisher's website www.larsen-keller.com

Table of Contents

Preface

The book aims to shed light on some of the unexplored aspects of agroecology. It discusses the various theories and concepts of this field in detail. Agroecology is an amalgamation of the ecological concepts and agricultural theories to produce better crop yield and its management. It strives to study a variety of ecosystems and is focused in the synthesis of effective agriculture practices and inclusive crop production techniques. It consists of the principles of all of these methods. This book is a valuable compilation of topics, ranging from the basic to the most complex theories and principles in the field of agroecology. Those in search of information to further their knowledge will be greatly assisted by this textbook.

A foreword of all Chapters of the book is provided below:

Chapter 1 - A number of ecological processes are applied to agricultural production systems. The study of these processes is known as agroecology. It is not affiliated with only one particular method of farming, and can be related with organic as well as integrated or conventional ways of farming. The chapter on agroecology offers an insightful focus, keeping in mind the complex subject matter.; **Chapter 2** - Agroecosystem analysis is an important part of the multidisciplinary subject known as agroecology. Agroecosystem considers aspects such as ecology, sociology and economics. Agrosystem analysis examines all these aspects and provides the reader with an in-depth understanding on topics related to agroecosystems. The chapter strategically encompasses and incorporates the major components and key concepts of agroecosystems, providing a complete understanding.; **Chapter 3** - Non-living things and living things both have effects on the environment. The negative effect that non-living things have on living things is termed as abiotic stress. Some of the examples of abiotic stressors include high winds, extreme temperature, drought and flood. The following text elucidates on topics such as natural stress and biotic stress, providing a complete understanding.; **Chapter 4** - The following section deals with the study of land-use sciences in agroecology. Some of these are forestry, agriculture, agroforestry, agronomy and polyculture. The cultivation of plants and animals is termed as agriculture whereas the science of preserving, managing and repairing forests is known as forestry. The topics discussed in the chapter are of great importance to broaden the existing knowledge on agroecology.; **Chapter 5** - The analysis of organisms and their environments is known as ecology. Ecologists seek to explain topics like life processes, adaptions and distribution of organisms. It is interdisciplinary and includes subjects like biology, geography and Earth science. This section is an overview of the subject matter incorporating all the major aspects of ecology, such as population ecology, restoration energy and ecological succession.; **Chapter 6** - Agroecology has a number of diverse applications. Some of these are organic farming, organic food, conservation agriculture, shifting cultivation and intercropping. Organic farming depends on fertilizers such as manure, compost and green manure. The food produced through the methods of organic farming is known ad organic food. This section serves as a source to understand the major categories related to agroecology.; **Chapter 7 -** Biodynamic agriculture is very similar to organic farming. The common features it shares with organic farming are its use of manures and composts. The methods that are unique to biodynamic agriculture are the approaches it takes for the treatment of crops and soil, and its emphasis on local production. This chapter will provide an integrated understanding of biodynamic agriculture.

I would like to thank the entire editorial team who made sincere efforts for this book and my family who supported me in my efforts of working on this book. I take this opportunity to thank all those who have been a guiding force throughout my life.

Editor

Introduction to Agroecology

A number of ecological processes are applied to agricultural production systems. The study of these processes is known as agroecology. It is not affiliated with only one particular method of farming, and can be related with organic as well as integrated or conventional ways of farming. The chapter on agroecology offers an insightful focus, keeping in mind the complex subject matter.

Agroecology

A community-supported agriculture share of crops

Agroecology is the study of ecological processes applied to agricultural production systems. The prefix *agro-* refers to *agriculture*. Bringing ecological principles to bear in agroecosystems can suggest novel management approaches that would not otherwise be considered. The term is often used imprecisely and may refer to "a science, a movement, [or] a practice." Agroecologists study a variety of agroecosystems, and the field of agroecology is not associated with any one particular method of farming, whether it be organic, integrated, or conventional; intensive or extensive. Although it has much more common thinking and principles with some of the before mentioned farming systems.

Ecological Strategy

Agroecologists do not unanimously oppose technology or inputs in agriculture but instead assess how, when, and if technology can be used in conjunction with natural, social and human assets. Agroecology proposes a context- or site-specific manner of studying agroecosystems, and as such, it recognizes that there is no universal formula or recipe for the success and maximum well-being of an agroecosystem. Thus, agroecology is not defined by certain management practices, such as the use of natural enemies in place of insecticides, or polyculture in place of monoculture.

Instead, agroecologists may study questions related to the four system properties of agroecosystems: productivity, stability, sustainability and equitability. As opposed to disciplines that are concerned with only one or some of these properties, agroecologists see all four properties as interconnected and integral to the success of an agroecosystem. Recognizing that these properties are found on varying spatial scales, agroecologists do not limit themselves to the study of agroecosystems at any one scale: gene-organism-population-community-ecosystem-landscape-biome, field-farm-community-region-state-country-continent-global.

Agroecologists study these four properties through an interdisciplinary lens, using natural sciences to understand elements of agroecosystems such as soil properties and plant-insect interactions, as well as using social sciences to understand the effects of farming practices on rural communities, economic constraints to developing new production methods, or cultural factors determining farming practices.

Approaches

Agroecologists do not always agree about what agroecology is or should be in the long-term. Different definitions of the term agroecology can be distinguished largely by the specificity with which one defines the term "ecology," as well as the term's potential political connotations. Definitions of agroecology, therefore, may be first grouped according to the specific contexts within which they situate agriculture. Agroecology is defined by the OECD as "the study of the relation of agricultural crops and environment." This definition refers to the "-ecology" part of "agroecology" narrowly as the natural environment. Following this definition, an agroecologist would study agriculture's various relationships with soil health, water quality, air quality, meso- and micro-fauna, surrounding flora, environmental toxins, and other environmental contexts.

A more common definition of the word can be taken from Dalgaard et al., who refer to agroecology as the study of the interactions between plants, animals, humans and the environment within agricultural systems. Consequently, agroecology is inherently multidisciplinary, including factors from agronomy, ecology, sociology, economics and related disciplines. In this case, the "-ecology" portion of "agroecology is defined broadly to include social, cultural, and economic contexts as well. Francis et al. also expand the definition in the same way, but put more emphasis on the notion of food systems.

Agroecology is also defined differently according to geographic location. In the global south, the term often carries overtly political connotations. Such political definitions of the term usually ascribe to it the goals of social and economic justice; special attention,

in this case, is often paid to the traditional farming knowledge of indigenous populations. North American and European uses of the term sometimes avoid the inclusion of
such overtly political goals. In these cases, agroecology is seen more strictly as a scientific discipline with less specific social goals.

Agro-population Ecology

This approach is derived from the science of ecology primarily based on population
ecology, which over the past three decades has been displacing the ecosystems biology
of Odum. Buttel explains the main difference between the two categories, saying that
"the application of population ecology to agroecology involves the primacy not only
of analyzing agroecosystems from the perspective of the population dynamics of their
constituent species, and their relationships to climate and biogeochemistry, but also
there is a major emphasis placed on the role of genetics."

Inclusive Agroecology

Rather than viewing agroecology as a subset of agriculture, Wojtkowski takes a more
encompassing perspective. In this, natural ecology and agroecology are the major
headings under ecology. Natural ecology is the study of organisms as they interact with
and within natural environments. Correspondingly, agroecology is the basis for the
land-use sciences. Here humans are the primary governing force for organisms within
planned and managed, mostly terrestrial, environments.

As key headings, natural ecology and agroecology provide the theoretical base for their
respective sciences. These theoretical bases overlap but differ in a major way. Economics has no role in the functioning of natural ecosystems whereas economics sets direction and purpose in agroecology.

Under agroecology are the three land-use sciences, agriculture, forestry, and agroforestry. Although these use their plant components in different ways, they share the same
theoretical core.

Beyond this, the land-use sciences further subdivide. The subheadings include agronomy,
organic farming, traditional agriculture, permaculture, and silviculture. Within this system
of subdivisions, agroecology is philosophically neutral. The importance lies in providing a
theoretical base hitherto lacking in the land-use sciences. This allows progress in biocom-
plex agroecosystems including the multi-species plantations of forestry and agroforestry.

Applications

To arrive at a point of view about a particular way of farming, an agroecologist would
first seek to understand the contexts in which the farm(s) is(are) involved. Each farm
may be inserted in a unique combination of factors or contexts. Each farmer may have
their own premises about the meanings of an agricultural endeavor, and these mean-

ings might be different from those of agroecologists. Generally, farmers seek a config-uration that is viable in multiple contexts, such as family, financial, technical, political, logistical, market, environmental, spiritual. Agroecologists want to understand the be-havior of those who seek livelihoods from plant and animal increase, acknowledging the organization and planning that is required to run a farm.

Views on Organic and Non-organic Milk Production

Because organic agriculture proclaims to sustain the health of soils, ecosystems, and people, it has much in common with Agroecology; this does not mean that Agroecology is synonymous with organic agriculture, nor that Agroecology views organic farming as the 'right' way of farming. Also, it is important to point out that there are large differ-ences in organic standards among countries and certifying agencies.

Three of the main areas that agroecologists would look at in farms, would be: the envi-ronmental impacts, animal welfare issues, and the social aspects.

Environmental impacts caused by organic and non-organic milk production can vary significantly. For both cases, there are positive and negative environmental conse-quences.

Compared to conventional milk production, organic milk production tends to have lower eutrophication potential per ton of milk or per hectare of farmland, because it potentially reduces leaching of nitrates (NO_3^-) and phosphates (PO_4^-) due to lower fertilizer appli-cation rates. Because organic milk production reduces pesticides utilization, it increases land use per ton of milk due to decreased crop yields per hectare. Mainly due to the lower level of concentrates given to cows in organic herds, organic dairy farms generally pro-duce less milk per cow than conventional dairy farms. Because of the increased use of roughage and the, on-average, lower milk production level per cow, some research has connected organic milk production with increases in the emission of methane.

Animal welfare issues vary among dairy farms and are not necessarily related to the way of producing milk (organically or conventionally).

A key component of animal welfare is freedom to perform their innate (natural) behav-ior, and this is stated in one of the basic principles of organic agriculture. Also, there are other aspects of animal welfare to be considered - such as freedom from hunger, thirst, discomfort, injury, fear, distress, disease and pain. Because organic standards require loose housing systems, adequate bedding, restrictions on the area of slatted floors, a minimum forage proportion in the ruminant diets, and tend to limit stocking densities both on pasture and in housing for dairy cows, they potentially promote good foot and hoof health. Some studies show lower incidence of placenta retention, milk fever, abo-masums displacement and other diseases in organic than in conventional dairy herds. However, the level of infections by parasites in organically managed herds is generally higher than in conventional herds.

Social aspects of dairy enterprises include life quality of farmers, of farm labor, of rural and urban communities, and also includes public health.

Both organic and non-organic farms can have good and bad implications for the life quality of all the different people involved in that food chain. Issues like labor conditions, labor hours and labor rights, for instance, do not depend on the organic/non-organic characteristic of the farm; they can be more related to the socio-economical and cultural situations in which the farm is inserted, instead.

As for the public health or food safety concern, organic foods are intended to be healthy, free of contaminations and free from agents that could cause human diseases. Organic milk is meant to have no chemical residues to consumers, and the restrictions on the use of antibiotics and chemicals in organic food production has the purpose to accomplish this goal. Although dairy cows in both organic and conventional farming practices can be exposed to pathogens, it has been shown that, because antibiotics are not permitted as a preventative measure in organic practices, there are far fewer antibiotic resistant pathogens on organic farms. This dramatically increases the efficacy of antibiotics when/if they are necessary.

In an organic dairy farm, an agroecologist could evaluate the following:

1. Can the farm minimize environmental impacts and increase its level of sustainability, for instance by efficiently increasing the productivity of the animals to minimize waste of feed and of land use?

2. Are there ways to improve the health status of the herd (in the case of organics, by using biological controls, for instance)?

3. Does this way of farming sustain good quality of life for the farmers, their families, rural labor and communities involved?

Views on No-till Farming

No-tillage is one of the components of conservation agriculture practices and is considered more environmental friendly than complete tillage. There is a general consensus that no-till can increase soils capacity of acting as a carbon sink, especially when combined with cover crops.

No-till can contribute to higher soil organic matter and organic carbon content in soils, though reports of no-effects of no-tillage in organic matter and organic carbon soil contents also exist, depending on environmental and crop conditions. In addition, no-till can indirectly reduce CO_2 emissions by decreasing the use of fossil fuels.

Most crops can benefit from the practice of no-till, but not all crops are suitable for complete no-till agriculture. Crops that do not perform well when competing with other plants that grow in untilled soil in their early stages can be best grown by using other conservation tillage practices, like a combination of strip-till with no-till areas. Also, crops which harvestable portion grows underground can have better results with strip-tillage, mainly in soils which are hard for plant roots to penetrate into deeper layers to access water and nutrients.

The benefits provided by no-tillage to predators may lead to larger predator populations, which is a good way to control pests (biological control), but also can facilitate predation of the crop itself. In corn crops, for instance, predation by caterpillars can be higher in no-till than in conventional tillage fields.

In places with rigorous winter, untilled soil can take longer to warm and dry in spring, which may delay planting to less ideal dates. Another factor to be considered is that organic residue from the prior year's crops lying on the surface of untilled fields can provide a favorable environment to pathogens, helping to increase the risk of transmitting diseases to the future crop. And because no-till farming provides good environment for pathogens, insects and weeds, it can lead farmers to a more intensive use of chemicals for pest control. Other disadvantages of no-till include underground rot, low soil temperatures and high moisture.

Based on the balance of these factors, and because each farm has different problems, agroecologists will not atest that only no-till or complete tillage is the right way of farming. Yet, these are not the only possible choices regarding soil preparation, since there are intermediate practices such as strip-till, mulch-till and ridge-till, all of them - just as no-till - categorized as conservation tillage. Agroecologists, then, will evaluate the need of different practices for the contexts in which each farm is inserted.

In a no-till system, an agroecologist could ask the following:

1. Can the farm minimize environmental impacts and increase its level of sustainability; for instance by efficiently increasing the productivity of the crops to minimize land use?

2. Does this way of farming sustain good quality of life for the farmers, their families, rural labor and rural communities involved?

History

Pre-WWII

The notions and ideas relating to crop ecology have been around since at least 1911 when F.H. King released *Farmers of Forty Centuries*. King was one of the pioneers as a proponent of more quantitative methods for characterization of water relations and physical properties of soils. In the late 1920s the attempt to merge agronomy and ecology was born with the development of the field of crop ecology. Crop ecology's main concern was where crops would be best grown. Actually, it was only in 1928 that agronomy and ecology were formally linked by Klages.

The first mention of the term agroecology was in 1928, with the publication of the term by Bensin in 1928. The book of Tischler (1965), was probably the first to be actually titled 'agroecology'. He analysed the different components (plants, animals, soils and climate) and their interactions within an agroecosystem as well as the impact of human agricultural management on these components. Other books dealing with agroecology, but without using the term explicitly were published by the German zoologist Friederichs (1930) with his book on agricultural zoology and related ecological/environmental factors for plant protection, and by American crop physiologist Hansen in 1939 when both used the word as a synonym for the application of ecology within agriculture.

Post-WWII

Gliessman mentions that post-WWII, groups of scientists with ecologists gave more focus to experiments in the natural environment, while agronomists dedicated their attention to the cultivated systems in agriculture. According to Gliessman, the two groups kept their research and interest apart until books and articles using the concept of agroecosystems and the word agroecology started to appear in 1970. Dalgaard explains the different points of view in ecology schools, and the fundamental differences, which set the basis for the development of agroecology. The early ecology school of Henry Gleason investigated plant populations focusing in the hierarchical levels of the organism under study.

Friederich Clement's ecology school, however included the organism in question as well as the higher hierarchical levels in its investigations, a "landscape perspective". However, the ecological schools where the roots of agroecology lie are even broader in nature. The ecology school of Tansley, whose view included both the biotic organism and their environment, is the one from which the concept of agroecosystems emerged in 1974 with Harper.

In the 1960s and 1970s the increasing awareness of how humans manage the landscape and its consequences set the stage for the necessary cross between agronomy and ecology. Even though, in many ways the environmental movement in the US was a product of the times, the Green Decade, spread an environmental awareness of the unintended consequences of changing ecological processes. Works such as *Silent Spring*, and *The Limits to Growth*, and changes in legislation such as the Clean Air Act, Clean Water Act, and the National Environmental Policy Act caused the public to be aware of societal growth patterns, agricultural production, and the overall capacity of the system.

Fusion with Ecology

After the 1970s, when agronomists saw the value of ecology and ecologists began to use the agricultural systems as study plots, studies in agroecology grew more rapidly. Gliessman describes that the innovative work of Prof. Efraim Hernandez X., who developed research based on indigenous systems of knowledge in Mexico, led to education programs in agroecology. In 1977 Prof. Efraim Hernandez X. explained that modern agricultural systems had lost their ecological foundation when socio-economic factors became the only driving force in the food system. The acknowledgement that the socio-economic interactions are indeed one of the fundamental components of any agro-ecosystems came to light in 1982, with the article Agroecologia del Tropico Americano by Montaldo. The author argues that the socio-economic context cannot be separated from the agricultural systems when designing agricultural practices.

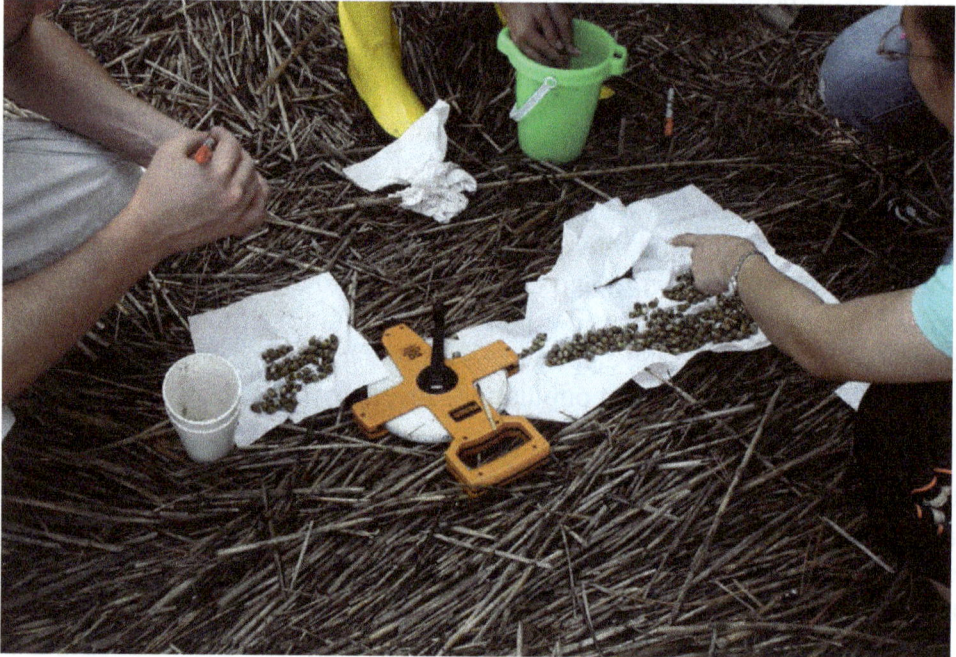

Fusion with ecology

In 1995 Edens et al. in Sustainable Agriculture and Integrated Farming Systems solidified this idea proving his point by devoting special sections to economics of the systems, ecological impacts, and ethics and values in agriculture. Actually, 1985 ended up being a fertile and creative year for the new discipline. For instance in the same year, Miguel Altieri integrated how consolidation of the farms, and cropping systems impact pest populations. In addition, Gliessman highlighted that socio-economic, technological, and ecological components give rise to producer choices of food production systems. These pioneering agroecologists have helped to frame the foundation of what we today consider the interdisciplinary field of agroecology and have led to advances in a number of farming systems. In Asian rice, for example, crop diversification by growing flowering crops in strips beside rice fields has recently been demonstrated to reduce pests so effectively (by the flower nectar attracting and supporting parasitoids and predators) that insecticide spraying is reduced by 70%, yields increase by 5%, together resulting in an economic advantage of 7.5%(Gurr et al., 2016).

By Region

The principles of agroecology are expressed differently depending on local ecological and social contexts.

Latin America

Latin America's experiences with North American Green Revolution agricultural techniques have opened space for agroecologists. Traditional or indigenous knowledge represents a wealth of possibility for agroecologists, including "exchange of wisdoms." See Miguel Alteiri's *Enhancing the Productivity of Latin American Traditional Peasant Farming Systems Through an Agroecological Approach* for information on agroecology in Latin America.

Madagascar

Most of the historical farming in Madagascar has been conducted by indigenous peoples. The French colonial period disturbed a very small percentage of land area, and even included some useful experiments in Sustainable forestry. Slash-and-burn techniques, a component of some shifting cultivation systems have been practised by natives in Madagascar for centuries. As of 2006 some of the major agricultural products from slash-and-burn methods are wood, charcoal and grass for Zebu grazing. These practices have taken perhaps the greatest toll on land fertility since the end of French rule, mainly due to overpopulation pressures.

Agrophysics

Agrophysics is a branch of science bordering on agronomy and physics, whose objects of study are the agroecosystem - the biological objects, biotope and biocoenosis affected by human activity, studied and described using the methods of physical sciences. Using the achievements of the exact sciences to solve major problems in agriculture, agrophysics involves the study of materials and processes occurring in the production and processing of agricultural crops, with particular emphasis on the condition of the environment and the quality of farming materials and food production.

Agrophysics is closely related to biophysics, but is restricted to the biology of the plants, animals, soil and an atmosphere involved in agricultural activities and biodiversity. It is different from biophysics in having the necessity of taking into account the specific features of biotope and biocoenosis, which involves the knowledge of nutritional science and agroecology, agricultural technology, biotechnology, genetics etc.

The needs of agriculture, concerning the past experience study of the local complex soil and next plant-atmosphere systems, lay at the root of the emergence of a new branch –

agrophysics – dealing this with experimental physics. The scope of the branch starting from soil science (physics) and originally limited to the study of relations within the soil environment, expanded over time onto influencing the properties of agricultural crops and produce as foods and raw postharvest materials, and onto the issues of quality, safety and labeling concerns, considered distinct from the field of nutrition for application in food science.

Research centres focused on the development of the agrophysical sciences include the Institute of Agrophysics, Polish Academy of Sciences in Lublin, and the Agrophysical Research Institute, Russian Academy of Sciences in St. Petersburg.

Agroecological Restoration

Agroecological restoration is the practice of re-integrating natural systems into agriculture in order to maximize sustainability, ecosystem services, and biodiversity. This is one example of a way to apply the principles of agroecology to an agricultural system.

Overview

Farms cannot be restored to a purely natural state because of the negative economic impact on farmers, but returning processes, such as pest control to nature with the method of intercropping, allows a farm to be more ecologically sustainable and, at the same time, economically viable. Agroecological restoration works toward this balance of sustainability and economic viability because conventional farming is not sustainable over the long run without the integration of natural systems and because the use of land for agriculture has been a driving force in creating the present world biodiversity crisis. Its efforts are complementary to, rather than a substitute for, biological conservation.

"...biodiversity is just as important on farms and in fields as it is in deep river valleys or mountain cloud forests."

FAO, 15 October 2004

Agriculture creates a conflict over the use of land between wildlife and humans. Though the domestication of crop plants occurred 10,000 years ago, a 500% increase in the amount of pasture and crop land over the last three hundred years has led to the rapid loss of natural habitats. In recent years, the world community acknowledged the value of biodiversity in treaties, such as the 1992 landmark Convention on Biological Diversity.

Reintegration

The reintegration of agricultural systems into more natural systems will result in decreased yield and produce a more complex system, but there will be considerable gains in biodiversity and ecosystem services.

Biodiversity

The Food and Agriculture Organization of the United Nations estimates that more than 40% of earth's land surface is currently used for agriculture. And because so much land has been converted to agriculture, habitat loss is recognized as the driving force in biodiversity loss (FAO). This biodiversity loss often occurred in two steps, as in the American Midwest, with the introduction of mixed farming carried out on small farms and then with the widespread use of mechanized farming and monoculture beginning after World War II. The decline in farmland biodiversity can now be traced to changes in farming practices and increased agricultural intensity.

Increasing Heterogeneity

Heterogeneity (here, the diversity or complexity of the landscape) has been shown to be associated with species diversity. For example, the abundance of butterflies has been found to increase with heterogeneity. One important part of maintaining heterogeneity in the spaces between different fields is made up of habitat that is not cropped, such as grass margins and strips, scrub along field boundaries, woodland, ponds, and fallow land. These seemingly unimportant pieces of land are crucial for the biodiversity of a farm. The presence of field margins benefits many different taxa: the plants attract herbivorous insects, will which attract certain species of birds and those birds will attract their natural predators. Also, the cover provided by the no cropped habitat allows the species that need a large range to move across the landscape.

Monoculture

In the absence of cover, species face a landscape in which their habitat is greatly fragmented. The isolation of a species to a small habitat that it can't safely wander from can create a genetic bottleneck, decreasing the resilience of the particular population, and be another factor leading to the decline of the total population of the species. Monoculture, the practice of producing a single crop over a wide area, causes fragmentation. In conventional farming, monoculture, such as with rotations of corn and soybean crops planted in alternating growing seasons, is used so that very high yields can be produced. After the mechanization of farming, monoculture became a standard practice in corn-beans rotation, and had broad implications for the long-term sustainability and biodiversity of farms. Whereas organic fertilizers, had kept the soil's nutrients fixed to the ecosystem, the introduction of monoculture removed the nutrients and farmers compensated for that loss by using inorganic fertilizers. It is estimated that humans have doubled the rate of nitrogen input into the nitrogen cycle, mostly since 1975. As a result, the biological processes that controlled the way crops used the nutrients changed and the leached nitrogen from farmland soils has become a source of pollution.

Organic Farming

Organic farming is defined in different legal terms by different nations, but its main distinction from conventional farming is that it prohibits the use of synthetic chemicals in crop and livestock production. Often, it also includes diverse crop rotations and provides non-cropped habitat for insects that provide ecosystem services, such as pest control and pollination. However, it is merely encouraged that organic farmers follow those kinds of wildlife friendly practices, and as a result there is a great difference between the ecosystem services that similarly sized but distinctly managed organic farms provide. A recent review of the 76 studies concerning the relationship between biodiversity and organic farming listed three practices associated with organic farming that accounted for the higher biodiversity counts found in organic farms as compared to conventional farms.

"1. *Prohibition/reduced use of chemical pesticides and inorganic fertilizers* is likely to have a positive impact through the removal of both direct and indirect negative effects on arable plants, invertebrates and vertebrates. 2. *Sympathetic management of non-crop habitats and field margins* can enhance diversity and abundance of arable plants, invertebrates, birds and mammals. 3. *Preservation of mixed farming* is likely to positively impact farmland biodiversity through the provision of greater habitat heterogeneity at a variety of temporal and spatial scales within the landscape."

References

- Encyclopedia of Soil Science, edts. Ward Chesworth, 2008, Uniw. of Guelph Canada, Publ. Springer, ISBN 978-1-4020-3994-2

- АГРОФИЗИКА - AGROPHYSICS by Е. В. Шеин (J.W. Chein), В. М. Гончаров (W.M. Gontcharow), Ростов-на-Дону (Rostov-on-Don), Феникс (Phoenix), 2006, - 399 с., ISBN 5-222-07741-1

- Soil Physical Condition and Plant Roots by J. Gliński, J. Lipiec, 1990, CRC Press, Inc., Boca Raton, USA, ISBN 0-8493-6498-1

- Soil Aeration and its Role for Plants by J. Gliński, W. Stępniewski, 1985, Publisher: CRC Press, Inc., Boca Raton, USA, ISBN 0-8493-5250-9

- Scientific Dictionary of Agrophysics: polish-English, polsko-angielski by R. Dębicki, J. Gliński, J. Horabik, R. T. Walczak - Lublin 2004, ISBN 83-87385-88-3

- Physical Methods in Agriculture. Approach to Precision and Quality, edts. J. Blahovec and M. Kutilek, Kluwer Academic Publishers, New York 2002, ISBN 0-306-47430-1.

- Encyclopedia of Agrophysics in series: Encyclopedia of Earth Sciences Series edts. Jan Glinski, Jozef Horabik, Jerzy Lipiec, 2011, Publisher: Springer, ISBN 978-90-481-3585-1

2

Agroecosystems and their Analysis

Agroecosystem analysis is an important part of the multidisciplinary subject known as agroecology. Agroecosystem considers aspects such as ecology, sociology and economics. Agrosystem analysis examines all these aspects and provides the reader with an in-depth understanding on topics related to agroecosystems. The chapter strategically encompasses and incorporates the major components and key concepts of agroecosystems, providing a complete understanding.

Agroecosystem

Agroecosystem in Croton-on-Hudson, New York in Westchester County.
Intercropped tomatoes, basil, peppers and eggplants.

An agroecosystem is the basic unit of study in agroecology, and is somewhat arbitrarily defined as a spatially and functionally coherent unit of agricultural activity, and includes the living and nonliving components involved in that unit as well as their interactions.

An agroecosystem can be viewed as a subset of a conventional ecosystem. As the name implies, at the core of an agroecosystem lies the human activity of agriculture. However, an agroecosystem is not restricted to the immediate site of agricultural activity (e.g. the farm), but rather includes the region that is impacted by this activity, usually by changes to the complexity of species assemblages and energy flows, as well as to the net nutrient balance. Traditionally an agroecosystem, particularly one managed intensively, is characterized as having a simpler species composition and simpler energy and nutrient flows than "natural" ecosystem. Likewise, agroecosystems are often associated with elevated nutrient input, much of which exits the farm leading to eutrophication of connected ecosystems not directly engaged in agriculture.

The Future for Farming?

Some major organizations are hailing farming within agroecosystems as the way forward for mainstream agriculture. Current farming methods have resulted in overstretched water resources, high levels of erosion and reduced soil fertility. According to a report by the International Water Management Institute and the United Nations Environment Programme, there is not enough water to continue farming using current practices; therefore how critical water, land, and ecosystem resources are used to boost crop yields must be reconsidered. The report suggested assigning value to ecosystems, recognizing environmental and livelihood tradeoffs, and balancing the rights of a variety of users and interests, as well addressing inequities that sometimes result when such measures are adopted, such as the reallocation of water from poor to rich, the clearing of land to make way for more productive farmland, or the preservation of a wetland system that limits fishing rights.

A Long Tradition

Forest gardens are probably the world's oldest and most resilient agroecosystem. Forest gardens originated in prehistoric times along jungle-clad river banks and in the wet foothills of monsoon regions. In the gradual process of a family improving their immediate environment, useful tree and vine species were identified, protected and improved whilst undesirable species were eliminated. Eventually superior foreign species were selected and incorporated into the family's garden.

One of the major efforts of disciplines such as agroecology is to promote management styles that blur the distinction between agroecosystems and "natural" ecosystems, both by decreasing the impact of agriculture (increasing the biological and trophic complexity of the agricultural system as well as decreasing the nutrient inputs/outflow) and by increasing awareness that "downstream" effects extend agroecosystems beyond the boundaries of the farm (e.g. the Corn Belt agroecosystem includes the hypoxic zone in the Gulf of Mexico). In the first case, polyculture or buffer strips for wildlife habitat can restore some complexity to a cropping system, while organic farming can reduce nutrient inputs. Efforts of the second type are most common at the watershed scale. An example is the National Association of Conservation Districts' Lake Mendota Watershed Project, which seeks to reduce runoff from the agricultural lands feeding into the lake with the aim of reducing algal blooms.

Agroecosystem Analysis

Agroecosystem analysis is a thorough analysis of an agricultural environment which considers aspects from ecology, sociology, economics, and politics with equal weight. There are many aspects to consider; however, it is literally impossible to account for all of them. This is one of the issues when trying to conduct an analysis of an agricultural environment. In the past, an agroecosystem analysis approach might be used to determine the sustainability of an agricultural system. It has become apparent, however, that the "sustainability" of the system depends heavily on the definition of sustainability chosen by the observer. Therefore, agroecosystem analysis is used to bring the richness of the true complexity of agricultural systems to an analysis to identify reconfigurations of the system (or *holon*) that will best suit individual situations.

Agroecosystem analysis is a tool of the multidisciplinary subject known as Agroecology. Agroecology and agroecosystem analysis are not the same as sustainable agriculture, though the use of agroecosystem analysis may help a farming system ensure its viability. Agroecosystem analysis is not a new practice, agriculturalists and farmers have been doing it since societies switched from hunting and gathering (hunter-gatherer) for food to settling in one area. Every time a person involved in agriculture evaluates their situation to identify methods to make the system function in a way that better suits their interests, they are performing an agroecosystem analysis.

Agroecosystem Analysis and Sustainable Agriculture Differ

It is difficult to discuss these differences without the aid of an example. Consider the case of a conventional apple farmer. This farmer may choose to change his farm to conform to the standards of USDA approved organic agriculture because he felt motivated by social or moral norms or the potential of increased profits or a host of other reasons. This farmer evaluated his situation and reconfigured it to try to improve it. Some might look at this situation and conclude that the apple farmer chose organic apple production because it is more sustainable for the environment. But, what if a few years later the farmer finds that he is struggling to make a profit and decides to go back to conventional agriculture? The farmer performed another agroecosystem analysis and arrived at a reconfiguration that some might see as unsustainable. This example illustrates how agroecosystem analysis is not required to lead a more environmentally sustainable form of agriculture. Agroecosystem analysis might produce a reconfiguration that is more economically sustainable or socially sustainable or politically sustainable for a farmer (or other actor). By definition, however, agroecosystem analysis is not required to produce an environmentally sustainable configuration for an agricultural system.

Approach to Analysis

William L. Bland, from the University of Wisconsin–Madison, developed the idea of a farm as a Holon (philosophy) This term, *holon*, was originally introduced by Arthur Koestler in 1966, in which he referred to a *holon* as an entity in which it is a part by itself, a *holon*, while contributing to a larger entity, which is also a *holon*. Bland develops this for an agricultural environment or farm as, "The farm holon is both the whole in which smaller holons exists, and a part of larger entities, themselves holons." This idea was expanded upon by Bland and Michael M. Bell University of Wisconsin–Madison in their 2007 article "A holon approach to agroecology," because it is difficult to account for boundary and change when using a systems thinking approach. One major difference between Koestler's holon and the holon idea developed for agroecosystem analysis is that the latter can only be defined as a holon if it has intentionality.

The farm itself is a holon and within the farm holon, other holons exist. For example, a farm animal, the farm family, and a farmworker can all be considered holons within the farm. Additionally, the farm is considered a holon which is inpart connected to other holons such as the county in which the farm resides, the bank from which the farmer borrowed money, or the grain elevator where the farmer can sell goods. Things like the tractor or the barn are not holons because they lack intentionality.

When conducting an agroecosystem analysis, the analyst should approach the farm as the farm itself and the "ecology of contexts" in which the farm and the farmer function. A "context" is anything that might influence functioning of the farm and cause it to change. According to Bland and Bell, examples of contexts include, "family, farm business, genetic heart disease, and spiritual beliefs." These examples illustrate the breadth of contexts that could influence why farmers do what they do. Bland concluded his model of a farm as a *holon* by stating, "A farm is not sustainable (disintegrates) when it cannot find an overall configuration that is simultaneously viable in all contexts."

Questions to Consider

There is no right or wrong way to evaluate an agroecosystem. It is important to identify all actors in a holon before beginning the analysis. When an analyst accepts the task of analyzing the agroecosystem, first and foremost, it must be approached as to incorporate all elements involved and should derive questions that should be answered. Questions such as:

- What defining factors (holons and contexts) determine the present configuration of the agroecosystem?

- How does one quantify the sustainability of the farm holon (economic, social, political, ecological and/or other)?

- How does the farmer or farm family perceive an agroecosystem?

- What is the farmer doing now, and how do those practices or actions affect the viability of the agroecosystem?

- Can the farmer maintain his livelihood continuing with current practices?

- What does the farmer value and where do those values come from?

- Will the farmer consider alternative farm configurations?

These are the types of questions an analyst could consider. There are no preset questions to ask, and usually more questions are derived than answered. However, the most important task an analysts can do, is to start the analysis with an open mind and under no presumptions about what is and is not sustainable for the farm holon.

Analysis Types

J. Visser of Dordt College uses a diagram, "Wealth Creation Wheel" to emphasize and account for the parameters of developing a thorough analysis. His diagram is more emphasized on economics; however, it is a useful tool to reference when starting to analyze an agroecoystem. His interest is to create a functioning wheel which will *roll* when all parameters are met equally. If one parameter is not functioning in context with the other parameters, then the wheel will be out of balance and ineffective, thus unsustainable. When referring to an agroecosystem, if one parameter is out of balance, this could lead to an unproductive cropping season and loss of income and/or livelihood.

References

- Koestler, Arthur, 1967. The Ghost in the Machine. London: Hutchinson. 1990 reprint edition, Penguin Group. ISBN 0-14-019192-5.

- Bland, W.L. and Bell, M.M., (2007) A holon approach to agroecology International Journal of Agricultural Sustainability 5(4), 280-294. abstract available here Archived April 26, 2012.

3

Major Stress in Agroecology

Non-living things and living things both have effects on the environment. The negative effect that non-living things have on living things is termed as abiotic stress. Some of the examples of abiotic stressors include high winds, extreme temperature, drought and flood. The following text elucidates on topics such as natural stress and biotic stress, providing a complete understanding.

Abiotic Stress

Abiotic stress is defined as the negative impact of non-living factors on the living organisms in a specific environment. The non-living variable must influence the environment beyond its normal range of variation to adversely affect the population performance or individual physiology of the organism in a significant way.

Whereas a biotic stress would include such living disturbances as fungi or harmful insects, abiotic stress factors, or stressors, are naturally occurring, often intangible, factors such as intense sunlight or wind that may cause harm to the plants and animals in the area affected. Abiotic stress is essentially unavoidable. Abiotic stress affects animals, but plants are especially dependent on environmental factors, so it is particularly constraining. Abiotic stress is the most harmful factor concerning the growth and productivity of crops worldwide. Research has also shown that abiotic stressors are at their most harmful when they occur together, in combinations of abiotic stress factors.

Examples

Abiotic stress comes in many forms. The most common of the stressors are the easiest for people to identify, but there are many other, less recognizable abiotic stress factors which affect environments constantly.

The most basic stressors include:

- High winds
- Extreme temperatures
- Drought
- Flood

- Other natural disasters, such as tornadoes and wildfires.

Lesser-known stressors generally occur on a smaller scale. They include: poor edaphic conditions like rock content and pH levels, high radiation, compaction, contamination, and other, highly specific conditions like rapid rehydration during seed germination.

Effects

Abiotic stress, as a natural part of every ecosystem, will affect organisms in a variety of ways. Although these effects may be either beneficial or detrimental, the location of the area is crucial in determining the extent of the impact that abiotic stress will have. The higher the latitude of the area affected, the greater the impact of abiotic stress will be on that area. So, a taiga or boreal forest is at the mercy of whatever abiotic stress factors may come along, while tropical zones are much less susceptible to such stressors.

Benefits

One example of a situation where abiotic stress plays a constructive role in an ecosystem is in natural wildfires. While they can be a human safety hazard, it is productive for these ecosystems to burn out every once in a while so that new organisms can begin to grow and thrive. Even though it is healthy for an ecosystem, a wildfire can still be considered an abiotic stressor, because it puts an obvious stress on individual organisms within the area. Every tree that is scorched and each bird nest that is devoured is a sign of the abiotic stress. On the larger scale, though, natural wildfires are positive manifestations of abiotic stress.

What also needs to be taken into account when looking for benefits of abiotic stress, is that one phenomenon may not affect an entire ecosystem in the same way. While a flood will kill most plants living low on the ground in a certain area, if there is rice there, it will thrive in the wet conditions. Another example of this is in phytoplankton and zooplankton. The same types of conditions are usually considered stressful for these two types of organisms. They act very similarly when exposed to ultraviolet light and most toxins, but at elevated temperatures the phytoplankton reacts negatively, while the thermophilic zooplankton reacts positively to the increase in temperature. The two may be living in the same environment, but an increase in temperature of the area would prove stressful only for one of the organisms.

Lastly, abiotic stress has enabled species to grow, develop, and evolve, furthering natural selection as it picks out the weakest of a group of organisms. Both plants and animals have evolved mechanisms allowing them to survive extremes.

Detriments

The most obvious detriment concerning abiotic stress involves farming. It has been claimed by one study that abiotic stress causes the most crop loss of any other factor and that most major crops are reduced in their yield by more than 50% from their potential yield.

Because abiotic stress is widely considered a detrimental effect, the research on this branch of the issue is extensive. For more information on the harmful effects of abiotic stress.

In Plants

A plant's first line of defense against abiotic stress is in its roots. If the soil holding the plant is healthy and biologically diverse, the plant will have a higher chance of surviving stressful conditions.

Facilitation, or the positive interactions between different species of plants, is an intricate web of association in a natural environment. It is how plants work together. In areas of high stress, the level of facilitation is especially high as well. This could possibly be because the plants need a stronger network to survive in a harsher environment, so their interactions between species, such as cross-pollination or mutualistic actions, become more common to cope with the severity of their habitat.

Plants also adapt very differently from one another, even from a plant living in the same area. When a group of different plant species was prompted by a variety of different stress signals, such as drought or cold, each plant responded uniquely. Hardly any of the responses were similar, even though the plants had become accustomed to exactly the same home environment.

Rice (*Oryza sativa*) is a classic example. Rice is a staple food throughout the world, especially in China and India. Rice plants experience different types of abiotic stresses, like drought and high salinity. These stress conditions have a negative impact on rice production. Genetic diversity has been studied among several rice varieties with different genotypes using molecular markers.

Serpentine soils (media with low concentrations of nutrients and high concentrations of heavy metals) can be a source of abiotic stress. Initially, the absorption of toxic metal ions is limited by cell membrane exclusion. Ions that are absorbed into tissues are sequestered in cell vacuoles. This sequestration mechanism is facilitated by proteins on the vacuole membrane.

Chemical priming has been proposed to increase tolerance to abiotic stresses in crop plants. In this method, which is analogous to vaccination, stress-inducing chemical agents are introduced to the plant in brief doses so that the plant begins preparing defense mechanisms. Thus, when the abiotic stress occurs, the plant has already prepared defense mechanisms that can be activated faster and increase tolerance.

Phosphate Starvation in Plants

Phosphorus (P) is an essential macronutrient required for plant growth and development, but most of the world's soil is limited in this important plant nutrient. Plants can utilize P mainly in the form if soluble inorganic phosphate (Pi) but are subjected to abiotic stress of P-limitation when there is not sufficient soluble PO_4 available in the soil. Phosphorus forms insoluble complexes with Ca and Mg in basic soils and Al and Fe in acidic soils that makes it unavailable for plant roots. When there is limited bioavailable P in the soil, plants show extensive abiotic stress phenotype such as short primary roots and more lateral roots and root hairs to make more surface available for Pi absorption, exudation of organic acids and phosphatase to release Pi from complex P containing molecules and make it available for growing plants organs. It has been shown that PHR1, a MYB - related transcription factor is a master regulator of P-starvation response in plants. PHR1 also has been shown to regulate extensive remodeling of lipids and metabolites during phosphorus limitation stress

In Animals

For animals, the most stressful of all the abiotic stressors is heat. This is because many species are unable to regulate their internal body temperature. Even in the species that are able to regulate their own temperature, it is not always a completely accurate system. Temperature determines metabolic rates, heart rates, and other very important factors within the bodies of animals, so an extreme temperature change can easily distress the animal's body. Animals can respond to extreme heat, for example, through natural heat acclimation or by burrowing into the ground to find a cooler space.

It is also possible to see in animals that a high genetic diversity is beneficial in providing resiliency against harsh abiotic stressors. This acts as a sort of stock room when a species is plagued by the perils of natural selection. A variety of galling insects are among the most specialized and diverse herbivores on the planet, and their extensive protections against abiotic stress factors have helped the insect in gaining that position of honor.

In Endangered Species

Biodiversity is determined by many things, and one of them is abiotic stress. If an environment is highly stressful, biodiversity tends to be low. If abiotic stress does not have a strong presence in an area, the biodiversity will be much higher.

This idea leads into the understanding of how abiotic stress and endangered species are related. It has been observed through a variety of environments that as the level of abiotic stress increases, the number of species decreases. This means that species are more likely to become population threatened, endangered, and even extinct, when and where abiotic stress is especially harsh.

Biotic Stress

Biotic stress is stress that occurs as a result of damage done to plants by other living organisms, such as bacteria, viruses, fungi, parasites, beneficial and harmful insects, weeds, and cultivated or native plants. Not to be confused with abiotic stress, which is the negative impact of non-living factors on the organisms in a specific environment such as sunlight, wind, salinity, over watering and drought. The types of biotic stresses imposed on a plant depend on both geography and climate and on the host plant and its ability to resist particular stresses. Although there are many kinds of biotic stress, the majority of plant diseases are caused by fungi. Biotic stress remains a broadly defined term and those who study it face many challenges, such as the greater difficulty in controlling biotic stresses in an experimental context compared to abiotic stress.

The damage caused by these various living and nonliving agents can appear very similar. Even with close observation, accurate diagnosis can be difficult. For example, browning of leaves on an oak tree caused by drought stress may appear similar to leaf browning caused by oak wilt, a serious vascular disease, or the browning cause by anthracnose, a fairly minor leaf disease.

Agriculture

It is a major focus of agricultural research, due to the vast economic losses caused by biotic stress to cash crops. The relationship between biotic stress and plant yield affects economic decisions as well as practical development. The impact of biotic injury on crop yield impacts population dynamics, plant-stressor coevolution, and ecosystem nutrient cycling.

Biotic stress also impacts horticultural plant health and natural habitats ecology. It also has dramatic changes in the host recipient.Plants are exposed to many stress factors, such as drought, high salinity or pathogens, which reduce the yield of the cultivated plants or affect the quality of the harvested products. *Arabidopsis thaliana* is often used as a model plant to study the responses of plants to different sources of stress.

In History

Biotic stresses have had huge repercussions for humanity; an example of this is the potato blight, an oomycete which caused widespread famine in England, Ireland and Belgium in the 1840s. Another example is grape phylloxera coming from North America in the 19th century, which led to the Great French Wine Blight.

Today

Losses to pests and disease in crop plants continue to pose a significant threat to agriculture and food security. During the latter half of the 20th century, agriculture became increasingly reliant on synthetic chemical pesticides to provide control of pests and diseases, especially within the intensive farming systems common in the developed world. However, in the 21st century, this reliance on chemical control is becoming unsustainable. Pesticides tend to have a limited lifespan due to the emergence of resistance in the target pests, and are increasingly recognised in many cases to have negative impacts on biodiversity, and on the health of agricultural workers and even consumers.

Tomorrow

Due to the implications of climate change, it is suspected that plants will have increased susceptibility to pathogens. Additionally, elevated threat of abiotic stresses (i.e. drought and heat) are likely to contribute to plant pathogen susceptibility.

Effect on Plant Growth

Photosynthesis

Many biotic stresses affect photosynthesis, as chewing insects reduce leaf area and virus infections reduce the rate of photosynthesis per leaf area. Vascular- wilt fungi compromise the water transport and photosynthesis by inducing stomata closure.

Response to Stress

Plants have co-evolved with their parasites for several hundred million years. This co-evolutionary process has resulted in the selection of a wide range of plant defences against microbial pathogens and herbivorous pests which act to minimise frequency and impact of attack. These defences include both physical and chemical adaptations, which may either be expressed constitutively, or in many cases, are activated only in response to attack. For example, utilization of high metal ion concentrations derived from the soil allow plants to reduce the harmful effects of biotic stressors (pathogens, herbivores etc.); meanwhile preventing the infliction of severe metal toxicity by way of safeguarding metal ion distribution throughout the plant with protective physiological pathways. Such induced resistance provides a mechanism whereby the costs of defence are avoided until defense is beneficial to the plant. At the same time, successful pests and pathogens have evolved mechanisms to overcome both constitutive and induced resistance in their particular host species. In order to fully understand and manipulate plant biotic stress resistance, we require a detailed knowledge of these interactions at a wide range of scales, from the molecular to the community level.

Cross Tolerance with Abiotic Stress

*   Evidence shows that a plant undergoing multiple stresses, both abiotic and biotic (usually pathogen or herbivore attack), can produce a positive effect on plant performance, by reducing their susceptibility to biotic stress compared to how they respond to individual stresses. The interaction leads to a crosstalk

between their respective hormone signalling pathways which will either induce or antagonize another restructuring genes machinery to increase tolerance of defense reactions.

- Reactive oxygen species (ROS) are key signalling molecules produced in response to biotic and abiotic stress cross tolerance. ROS are produced in response to biotic stresses during the oxidative burst.

- Dual stress imposed by ozone (O3) and pathogen affects tolerance of crop and leads to altered host pathogen interaction (Fuhrer, 2003). Alteration in pathogenesis potential of pest due to O3 exposure is of ecological and economical importance.

- Tolerance to both biotic and abiotic stresses has been achieved. In maize, breeding programmes have led to plants which are tolerant to drought and have additional resistance to the parasitic weed *Striga hermonthica*.

Remote Sensing

The Agricultural Research Service (ARS) and various government agencies and private institutions have provided a great deal of fundamental information relating spectral reflectance and thermal emittance properties of soils and crops to their agronomic and biophysical characteristics. This knowledge has facilitated the development and use of various remote sensing methods for non-destructive monitoring of plant growth and development and for the detection of many environmental stresses that limit plant productivity. Coupled with rapid advances in computing and position locating technologies, remote sensing from ground-, air-, and space-based platforms is now capable of providing detailed spatial and temporal information on plant response to their local environment that is needed for site specific agricultural management approaches. This is very important in today's society because with increasing pressure on global food productivity due to population increase, result in a demand for stress-tolerant crop varieties that has never been greater.

Natural Stress

In regard to agriculture, Abiotic stress is stress produced by natural environment factors such as extreme temperatures, wind, drought, and salinity. Humankind doesn't have much control over abiotic stresses. It is very important for humans to understand how stress factors affect plants and other living things so that we can take some preventative measures.

Preventative measures are the only way that humans can protect themselves and their possessions from abiotic stress. There are many different types of abiotic stressors, and several methods that humans can use to reduce the negative effects of stress on living things.

Cold

One of the types of Abiotic Stress is cold. This has a huge impact on farmers. Cold impacts crop growers all over the world in every single country. Yields suffer and farmers also suffer huge losses because the weather is just too cold to produce crops (Xiong & Zhu, 2001).

Humans have planned the planting of our crops around the seasons. Even though the seasons are fairly predictable, there are always unexpected storms, heat waves, or cold snaps that can ruin our growing seasons.(Suzuki & Mittler, 2006)

ROS stands for reactive oxygen species. ROS plays a large role in mediating events through transduction. Cold stress was shown to enhance the transcript, protein, and activity of different ROS-scavenging enzymes. Low temperature stress has also been shown to increase the H_2O_2 accumulation in cells.(Suzuki & Mittler, 2006)

Plants can be acclimated to low or even freezing temperatures. If a plant can go through a mild cold spell this activates the cold-responsive genes in the plant. Then if the temperature drops again, the genes will have conditioned the plant to cope with the low temperature. Even below freezing temperatures can be survived if the proper genes are activated (Suzuki & Mittler, 2006).

Heat

Heat stress has been shown to cause problems in mitochondrial functions and can result in oxidative damage. Activators of heat stress receptors and defenses are thought to be related to ROS. Heat is another thing that plants can deal with if they have the proper pretreatment. This means that if the temperature gradually warms up the plants are going to be better able to cope with the change. A sudden long temperature increase could cause damage to the plant because their cells and receptors haven't had enough time to prepare for a major temperature change.

Heat stress can also have a detrimental effect on plant reproduction. Temperatures 10 degrees Celsius or more above normal growing temperatures can have a bad effect on several plant reproductive functions. Pollen meiosis, pollen germination, ovule development, ovule viability, development of the embryo, and seedling growth are all aspects of plant reproduction that are affected by heat.(Cross, McKay, McHughen, & Bonham-Smith, 2003)

There have been many studies on the effects of heat on plant reproduction. One study on plants was conducted on Canola plants at 28 degrees Celsius, the result was decreased plant size, but the plants were still fertile. Another experiment was conducted on Canola plants at 32 degrees Celsius, this resulted in the production of sterile plants. Plants seem to be more easily damaged by extreme temperatures during the late flower to early seed development stage (Cross, McKay, McHughen, & Bonham-Smith, 2003).

Wind

Wind is a huge part of abiotic stress. There is simply no way to stop the wind from blowing. This is definitely a bigger problem in some parts of the world than in others. Barren areas such as deserts are very susceptible to natural wind erosion. These types of areas don't have any vegetation to hold the soil particles in place. Once the wind starts to blow the soil around, there is nothing to stop the process. The only chance for the soil to stay in place is if the wind doesn't blow. This is usually not an option.

Plant growth in windblown areas is very limited. Because the soil is constantly moving, there is no opportunity for plants to develop a root system. Soil that blows a lot usually is very dry also. This leaves little nutrients to promote plant growth.

Farmland is typically very susceptible to wind erosion. Most farmers do not plant cover crops during the seasons when their main crops are not in the fields. They simply leave the ground open and uncovered. When the soil is dry, the top layer becomes similar to powder. When the wind blows, the powdery top layer of the farmland is picked up and carried for miles. This is the exact scenario that occurred during the "Dust Bowl" in the 30's. The combination of drought and poor farming practices allowed the wind to moves thousands of tons of dirt from one area to the next.

Wind is one of the factors that humans can really have some control over. Simply practice good farming practices. Don't leave ground bare and without any type of vegetation. During dry seasons it is especially important to have the land covered because dry soil moves much easier than wet soil in the wind.

When soil is not blowing due to the wind, conditions are much better for plant growth. Plants cannot grow in a soil that is constantly blowing. Their root systems do not have time to be established. Also, when soil particles are blowing they wear away at the plants that they run into. Plants are essentially "sand blasted."

Drought

Drought is very detrimental to all types of plant growth. When there is no water in the soil there are not very many nutrients to support plant growth. Drought also enhances the effects of wind. When drought occurs the soil becomes very dry and light. The wind picks up this dry dirt and carries it away. This action severely degrades the soil and creates a poor condition for growing plants.

Adaptation of Plants

Plants have been exposed to the elements for thousands of years. During this time they have evolved in order to lessen the effects of abiotic stress. Signal transduction is the mechanism in plants that is responsible for the adaptation of plants (Xiong & Zhu, 2001). Many signaling transduction networks have been discovered and studied in microbial and animal systems. There is limited knowledge in the plant field because it is very difficult to find exactly which phenotypes in the plant are affected by stressors. These phenotypes are very valuable to the researchers. They need to know the pheno-types so that they can create a method to screen for mutant genes. Mutants are the key to finding signaling pathways in living creatures.

Animals and microbes easier to run tests on because they show a reaction fairly quickly when a stress factor is put on them, this leads to the isolation of the specific gene. There have been decades of research on the effects of temperature, drought, and salinity, but not very many answers.

Receptors

The part of the plants, animal, or microbe that first senses an abiotic stress factor is a receptor. Once a signal is picked up by a receptor, a lot of different things can happen. Signals are transmitted intercellularly and then they activate nuclear transcription to get the effects of a certain set of genes. These genes that are activated allow the plant to respond to the stress that it is experiencing. Even though none of the receptors for cold, drought, salinity or the stress hormone abscisic acid in plants is known for sure, the knowledge that we have today shows that receptor-like protein kinases, two-com-ponent histidine kinases, as well as G-protein receptors may be the possible sensors of these different signals.

Receptor like kinases can be found in plants as well as animals. There are many more RLKs in plants than there are in animals. They are also a little bit different. Unlike ani-mal RLKs that usually possess tyrosine signature sequences, plant RLKs have serine or threonine signature sequences (Xiong & Zhu, 2001).

Genetically Modified Plants

Plants are most commonly modified to be resistant to specific herbicides or pathogens, but we have the technology to modify plants in order to make them resistant to specific abiotic stressors. Cold, heat, drought, or salt are all factors that could possibly be de-fended against by genetically modified plants.

Some plants could have genes added to them from other species of plants that have a resistance to a specific stress. Plants implanted with these genes would then become transgenic plants because they have the genes from another species of plant in them. Scientists first have to isolate the specific gene in a plant that is responsible for its resis-tance. The gene would then be taken out of the plant and put in to another plant. The plant that is injected with the new resistant gene would have a resistance to an abiotic stressor and be able to tolerate a wider range of conditions (Weil, 2005).

This process of creating transgenic plants could have a huge impact on our nation's economy. If plants could be genetically engineered to be resistant to a wider variety of stress, crop yields would skyrocket. With the expansion of town and cities there is a decreasing number of farm acres. Although the farm acres are being built on, the number of people consuming agriculture products is going up. Ethanol is also responsible for using much more of the corn that is grown here in the U.S. The production of this fuel has put a strain on the corn market. Prices of corn have gone up and this price is having a negative impact on the people who feed animals using corn. The combination of reduced acres of farmland and a higher demand on crops have left producers and consumers in a severe dilemma. The only solution to this problem is to keep getting higher and higher yields from the cropland that we have left.

Genetically modified plants are a good answer to the problem of not enough crops to go around. These plants can be engineered to be resistant to all types of abiotic stress. This would eliminate crop yield loss due to extreme temperatures, drought, wind, or salinity. The consumers of crops would enjoy a little bit lower prices because the demand on them would be a little lower.

The Midwestern U.S. is experiencing a severe drought. Farmers are being limited on how much they can irrigate due to the shortage of water. There is also very little rain during the growing season so the crops do not yield very well. This problem could be solved by genetically modifying plants to become more drought resistant. If plants could use less water and produce yields that are superior or equal to current ones, it would be better for the people and also the environment. People would enjoy an abundance of crops to consume and export for a profit. The environment would be able to have more water in its aquifers and rivers throughout the country.

Another environmental factor that would be improved would be the amount of land left for wildlife. Crops modified to be resistant to abiotic stress and other factors that decrease yields would require less land use. Producers would be able to grow enough crops on less acres if the plants were modified to produce very high yields. This would allow some of the cropland that is in use today to be set aside for wildlife. Instead of farming "fence line to fence line" farmers would be able to create large buffers in their fields. These buffers would provide a great habitat for plants and animals.

A lot of people do not like genetically modified organisms. People opposed to these modified plants often claim that they are not safe for the environment or for human consumption. There are many videos and reports in circulation that discredit the safety of genetically modified organisms. King Corn is one video that claims that corn is bad for humans to consume.

There are strict regulations and protocols that go along with genetically modifying plants. A company that specializes in producing genetically modifying organisms must put their plants through a huge variety of tests to ensure the safety of their product. Each of these tests must be passed by the product in order to produce more of the plant seeds.

When seeds are mass-produced, the fields that they are grown in have to meet specific criteria. They must have no vegetation zones around them to prevent the spread of the modified plants into the native population. The plots must be carefully labeled and marked so that the company knows exactly what is planted in the field. All of these protocols are in place to ensure the safety of the consumers and also of the environment. Because genetically modified plants are given stress resistant genes or high yielding genes they are better for the environment. They only help create more land to be put back into natural habitats for plants and animals.

Conclusion

onAbiotic stress is a naturally occurring factor that cannot be controlled by humans. Some of the stress factors go hand in hand. One example of two stressors that are complimentary to each other is wind and draught. Drought dries out the soil and kills the plants that are growing in the soil. After this occurs the soil is left barren and dry. When the wind picks up then the soil is picked up and carried for miles. Irrigation is a way that humans can try to keep this from happening, but sometimes it is not possible to irrigate some areas.

Genetically modified plants can be implemented to slow down the effects of the abiotic stressors. These plants can be given genes that allow them to survive several types of natural stressors. This allows more crops to be grown on a smaller amount of land. This is important because there is less and less farmland available. Also, less need for farmland allows some of it to be set aside for natural wildlife habitat.

Abiotic stress only poses a problem to people or the environment if they are not prepared for it. There can be steps taken by humans to lessen the effects. Plants and animals have the ability to adapt to abiotic stress over time. This is natures way of taking care of itself and keeping everything in balance.

4

Study of Land-use Sciences in Agroecology

The following section deals with the study of land-use sciences in agroecology. Some of these are forestry, agriculture, agroforestry, agronomy and polyculture. The cultivation of plants and animals is termed as agriculture whereas the science of preserving, managing and repairing forests is known as forestry. The topics discussed in the chapter are of great importance to broaden the existing knowledge on agroecology.

Agriculture

Fields in Záhorie (Slovakia) – a typical Central European agricultural region

Agriculture is the cultivation of animals, plants and fungi for food, fiber, biofuel, medicinal plants and other products used to sustain and enhance human life. Agriculture was the key development in the rise of sedentary human civilization, whereby farming of domesticated species created food surpluses that nurtured the development of civilization. The study of agriculture is known as agricultural science. The history of agriculture dates back thousands of years, and its development has been driven and defined by greatly different climates, cultures, and technologies. Industrial agriculture based on large-scale monoculture farming has become the dominant agricultural methodology.

Modern agronomy, plant breeding, agrochemicals such as pesticides and fertilizers, and technological developments have in many cases sharply increased yields from cultivation, but at the same time have caused widespread ecological damage and negative human health effects. Selective breeding and modern practices in animal husbandry have similarly increased the output of meat, but have raised concerns about animal welfare and the health effects of the antibiotics, growth hormones, and other chemicals commonly used in industrial meat production. Genetically modified organisms are an increasing component of agriculture, although they are banned in several countries. Agricultural food production and water management are increasingly becoming global issues that are fostering debate on a number of fronts. Significant degradation of land and water resources, including the depletion of aquifers, has been observed in recent decades, and the effects of global warming on agriculture and of agriculture on global warming are still not fully understood.

Domestic sheep and a cow (heifer) pastured together in South Africa

The major agricultural products can be broadly grouped into foods, fibers, fuels, and raw materials. Specific foods include cereals (grains), vegetables, fruits, oils, meats and spices. Fibers include cotton, wool, hemp, silk and flax. Raw materials include lumber and bamboo. Other useful materials are also produced by plants, such as resins, dyes, drugs, perfumes, biofuels and ornamental products such as cut flowers and nursery plants. Over one third of the world's workers are employed in agriculture, second only to the service sector, although the percentages of agricultural workers in developed countries has decreased significantly over the past several centuries.

Etymology and Terminology

The word *agriculture* is a late Middle English adaptation of Latin *agricultūra*, from *ager*, "field", and *cultūra*, "cultivation" or "growing". Agriculture usually refers to human activities, although it is also observed in certain species of ant, termite and ambrosia beetle. To practice agriculture means to use natural resources to "produce commodities which maintain life, including food, fiber, forest products, horticultural crops, and their related services." This definition includes arable farming or agronomy, and horticulture, all terms for the growing of plants, animal husbandry and forestry. A distinction is sometimes made between forestry and agriculture, based on the former's longer management rotations, extensive versus intensive management practices and development mainly by nature, rather than by man. Even then, it is acknowledged that there is a large amount of knowledge transfer and overlap between silviculture (the management of forests) and agriculture. In traditional farming, the two are often combined even on small landholdings, leading to the term agroforestry.

History

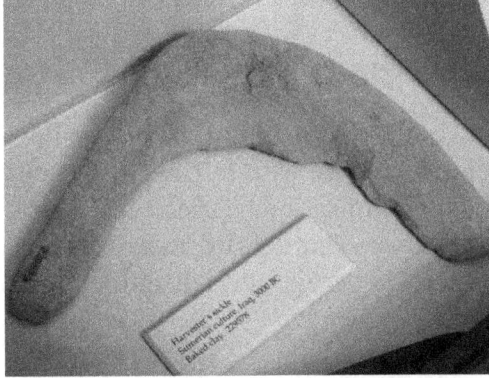

A Sumerian harvester's sickle made from baked clay (c. 3000 BC)

Agriculture began independently in different parts of the globe, and included a diverse range of taxa. At least 11 separate regions of the Old and New World were involved as independent centers of origin. Wild grains were collected and eaten from at least 105,000 years ago. Pigs were domesticated in Mesopotamia around 15,000 years ago. Rice was domesticated in China between 13,500 and 8,200 years ago, followed by mung, soy and azuki beans. Sheep were domesticated in Mesopotamia between 13,000 and 11,000 years ago. From around 11,500 years ago, the eight Neolithic founder crops, emmer and einkorn wheat, hulled barley, peas, lentils, bitter vetch, chick peas and flax were cultivated in the Levant. Cattle were domesticated from the wild aurochs in the areas of modern Turkey and Pakistan some 10,500 years ago. In the Andes of South America, the potato was domesticated between 10,000 and 7,000 years ago, along with beans, coca, llamas, alpacas, and guinea pigs. Sugarcane and some root vegetables were domesticated in New Guinea around 9,000 years ago. Sorghum was domesticated in the Sahel region of Africa by 7,000 years ago. Cotton was domesticated in Peru by 5,600 years ago, and was independently domesticated in Eurasia at an unknown time. In Mesoamerica, wild teosinte was domesticated to maize by 6,000 years ago.

In the Middle Ages, both in the Islamic world and in Europe, agriculture was transformed with improved techniques and the diffusion of crop plants, including the introduction of sugar, rice, cotton and fruit trees such as the orange to Europe by way of Al-Andalus. After 1492, the Columbian exchange brought New World crops such as maize, potatoes, sweet potatoes and manioc to Europe, and Old World crops such as wheat, barley, rice and turnips, and livestock including horses, cattle, sheep and goats to the Americas. Irrigation, crop rotation, and fertilizers were introduced soon after the Neolithic Revolution and developed much further in the past 200 years, starting with the British Agricultural Revolution. Since 1900, agriculture in the developed nations, and to a lesser extent in the developing world, has seen large rises in productivity as human labor has been replaced by mechanization, and assisted by synthetic fertilizers, pesticides, and selective breeding. The Haber-Bosch method allowed the synthesis of ammonium nitrate fertilizer on an industrial scale, greatly increasing crop yields. Modern agriculture has raised political issues including water pollution, biofuels, genetically modified organisms, tariffs and farm subsidies, leading to alternative approaches such as the organic movement.

Agriculture and Civilization

Civilization was the product of the Agricultural Neolithic Revolution. In the course of history, civilization coincided in space with fertile areas such as The Fertile Crescent, and states formed mainly in circumscribed agricultural lands. The Great Wall of China and the Roman empire's *limes* (borders) demarcated the same northern frontier of cereal agriculture. This cereal belt fed the civilizations formed in the Axial Age and connected by the Silk Road.

Ancient Egyptians, whose agriculture depended exclusively on the Nile, deified the river, worshipped, and exalted it in a great hymn. The Chinese imperial court issued numerous edicts, stating: "Agriculture is the foundation of this Empire." Egyptian, Mesopotamian, Chinese, and Inca Emperors themselves plowed ceremonial fields in order to show personal example to everyone.

Ancient strategists, Chinese Guan Zhong and Shang Yang and Indian Kautilya, drew doctrines linking agriculture with military power. Agriculture defined the limits on how large and for how long an army could be mobilized. Shang Yang called agriculture and war the *One*. In the vast human pantheon of agricultural deities there are several deities who combined the functions of agriculture and war.

As the Neolithic Agricultural Revolution produced civilization, the modern Agricultural Revolution, begun in Britain (British Agricultural Revolution), made possible the Industrial civilization. The first precondition for industry was greater yields by less manpower, resulting in greater percentage of manpower available for non-agricultural sectors.

Types of Agriculture

Reindeer herds form the basis of pastoral agriculture for several Arctic and Subarctic peoples.

Pastoralism involves managing domesticated animals. In nomadic pastoralism, herds of livestock are moved from place to place in search of pasture, fodder, and water. This type of farming is practised in arid and semi-arid regions of Sahara, Central Asia and some parts of India.

In shifting cultivation, a small area of a forest is cleared by cutting down all the trees and the area is burned. The land is then used for growing crops for several years. When the soil becomes less fertile, the area is then abandoned. Another patch of land is selected and the process is repeated. This type of farming is practiced mainly in areas with abundant rainfall where the forest regenerates quickly. This practice is used in Northeast India, Southeast Asia, and the Amazon Basin.

Subsistence farming is practiced to satisfy family or local needs alone, with little left over for transport elsewhere. It is intensively practiced in Monsoon Asia and South-East Asia.

In intensive farming, the crops are cultivated for commercial purpose i.e., for selling. The main motive of the farmer is to make profit, with a low fallow ratio and a high use of inputs. This type of farming is mainly practiced in highly developed countries.

Contemporary Agriculture

Satellite image of farming in Minnesota

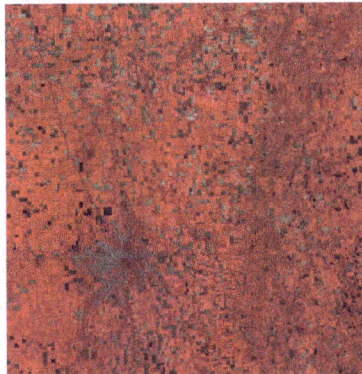

Infrared image of the above farms. Various colors indicate healthy crops (red),
flooding (black) and unwanted pesticides (brown).

In the past century, agriculture has been characterized by increased productivity, the substitution of synthetic fertilizers and pesticides for labor, water pollution, and farm subsidies. In recent years there has been a backlash against the external environmental effects of conventional agriculture, resulting in the organic and sustainable agriculture movements. One of the major forces behind this movement has been the European Union, which first certified organic food in 1991 and began reform of its Common Agricultural Policy (CAP) in 2005 to phase out commodity-linked farm subsidies, also known as decoupling. The growth of organic farming has renewed research in alternative technologies such as integrated pest management and selective breeding. Recent mainstream technological developments include genetically modified food.

In 2007, higher incentives for farmers to grow non-food biofuel crops combined with other factors, such as over development of former farm lands, rising transportation costs, climate change, growing consumer demand in China and India, and population growth, caused food shortages in Asia, the Middle East, Africa, and Mexico, as well as rising food prices around the globe. As of December 2007, 37 countries faced food crises, and 20 had imposed some sort of food-price controls. Some of these shortages resulted in food riots and even deadly stampedes. The International Fund for Agricultural Development posits that an increase in smallholder agriculture may be part of the solution to concerns about food prices and overall food security. They in part base this on the experience of Vietnam, which went from a food importer to large food exporter and saw a significant drop in poverty, due mainly to the development of smallholder agriculture in the country.

Disease and land degradation are two of the major concerns in agriculture today. For example, an epidemic of stem rust on wheat caused by the Ug99 lineage is currently spreading across Africa and into Asia and is causing major concerns due to crop losses of 70% or more under some conditions. Approximately 40% of the world's agricultural land is seriously degraded. In Africa, if current trends of soil degradation continue, the continent might be able to feed just 25% of its population by 2025, according to United Nations University's Ghana-based Institute for Natural Resources in Africa.

Agrarian structure is a long-term structure in the Braudelian understanding of the concept. On a larger scale the agrarian structure is more dependent on the regional, social, cultural and historical factors than on the state's undertaken activities. Like in Poland, where despite running an intense agrarian policy for many years, the agrarian structure in 2002 has much in common with that found in 1921 soon after the partitions period.

In 2009, the agricultural output of China was the largest in the world, followed by the European Union, India and the United States, according to the International Monetary Fund. Economists measure the total factor productivity of agriculture and by this measure agriculture in the United States is roughly 1.7 times more productive than it was in 1948.

Workforce

As of 2011, the International Labour Organization states that approximately one billion people, or over 1/3 of the available work force, are employed in the global agricultural sector. Agriculture constitutes approximately 70% of the global employment of children, and in many countries employs the largest percentage of women of any industry. The service sector only overtook the agricultural sector as the largest global employer in 2007. Between 1997 and 2007, the percentage of people employed in agriculture fell by over four percentage points, a trend that is expected to continue. The number of people employed in agriculture varies widely on a per-country basis, ranging from less than 2% in countries like the US and Canada to over 80% in many African nations. In developed countries, these figures are significantly lower than in previous centuries. During the 16th century in Europe, for example, between 55 and 75 percent of the population was engaged in agriculture, depending on the country. By the 19th century in Europe, this had dropped to between 35 and 65 percent. In the same countries today, the figure is less than 10%.

Safety

Rollover protection bar on a Fordson tractor

Agriculture, specifically farming, remains a hazardous industry, and farmers worldwide remain at high risk of work-related injuries, lung disease, noise-induced hearing loss, skin diseases, as well as certain cancers related to chemical use and prolonged sun exposure. On industrialized farms, injuries frequently involve the use of agricultural machinery, and a common cause of fatal agricultural injuries in developed countries is tractor rollovers. Pesticides and other chemicals used in farming can also be hazardous to worker health, and workers exposed to pesticides may experience illness or have children with birth defects. As an industry in which families commonly share in work and live on the farm itself, entire families can be at risk for injuries, illness, and death. Common causes of fatal injuries among young farm workers include drowning, machinery and motor vehicle-related accidents.

The International Labour Organization considers agriculture "one of the most hazard-ous of all economic sectors." It estimates that the annual work-related death toll among agricultural employees is at least 170,000, twice the average rate of other jobs. In ad-dition, incidences of death, injury and illness related to agricultural activities often go unreported. The organization has developed the Safety and Health in Agriculture Convention, 2001, which covers the range of risks in the agriculture occupation, the prevention of these risks and the role that individuals and organizations engaged in agriculture should play.

Agricultural Production Systems

Crop Cultivation Systems

Rice cultivation in Bihar, India

Cropping systems vary among farms depending on the available resources and con-straints; geography and climate of the farm; government policy; economic, social and political pressures; and the philosophy and culture of the farmer.

Shifting cultivation (or slash and burn) is a system in which forests are burnt, releas-ing nutrients to support cultivation of annual and then perennial crops for a period of several years. Then the plot is left fallow to regrow forest, and the farmer moves to a new plot, returning after many more years (10 – 20). This fallow period is shortened if population density grows, requiring the input of nutrients (fertilizer or manure) and some manual pest control. Annual cultivation is the next phase of intensity in which there is no fallow period. This requires even greater nutrient and pest control inputs.

The Banaue Rice Terraces in Ifugao, Philippines

Further industrialization led to the use of monocultures, when one cultivar is planted on a large acreage. Because of the low biodiversity, nutrient use is uniform and pests tend to build up, necessitating the greater use of pesticides and fertilizers. Multiple cropping, in which several crops are grown sequentially in one year, and intercropping, when several crops are grown at the same time, are other kinds of annual cropping systems known as polycultures.

In subtropical and arid environments, the timing and extent of agriculture may be limited by rainfall, either not allowing multiple annual crops in a year, or requiring irrigation. In all of these environments perennial crops are grown (coffee, chocolate) and systems are practiced such as agroforestry. In temperate environments, where ecosystems were predominantly grassland or prairie, highly productive annual farming is the dominant agricultural system.

Crop Statistics

Important categories of crops include cereals and pseudocereals, pulses (legumes), forage, and fruits and vegetables. Specific crops are cultivated in distinct growing regions throughout the world. In millions of metric tons, based on FAO estimate.

Livestock Production Systems

Ploughing rice paddy fields with water buffalo, in Indonesia

Animals, including horses, mules, oxen, water buffalo, camels, llamas, alpacas, donkeys, and dogs, are often used to help cultivate fields, harvest crops, wrangle other animals, and transport farm products to buyers. Animal husbandry not only refers to the breeding and raising of animals for meat or to harvest animal products (like milk, eggs, or wool) on a continual basis, but also to the breeding and care of species for work and companionship.

Livestock production systems can be defined based on feed source, as grassland-based, mixed, and landless. As of 2010, 30% of Earth's ice- and water-free area was used for producing livestock, with the sector employing approximately 1.3 billion people. Between the 1960s and the 2000s, there was a significant increase in livestock production, both by numbers and by carcass weight, especially among beef, pigs and chickens, the latter of which had production increased by almost a factor of 10. Non-meat animals, such as milk cows and egg-producing chickens, also showed significant production increases. Global cattle, sheep and goat populations are expected to continue to increase sharply through 2050. Aquaculture or fish farming, the production of fish for human consumption in confined operations, is one of the fastest growing sectors of food production, growing at an average of 9% a year between 1975 and 2007.

Oxen driven ploughs in India

During the second half of the 20th century, producers using selective breeding focused on creating livestock breeds and crossbreeds that increased production, while mostly disregarding the need to preserve genetic diversity. This trend has led to a significant decrease in genetic diversity and resources among livestock breeds, leading to a corresponding decrease in disease resistance and local adaptations previously found among traditional breeds.

Grassland based livestock production relies upon plant material such as shrubland, rangeland, and pastures for feeding ruminant animals. Outside nutrient inputs may be used, however manure is returned directly to the grassland as a major nutrient source. This system is particularly important in areas where crop production is not feasible because of climate or soil, representing 30 – 40 million pastoralists. Mixed production systems use grassland, fodder crops and grain feed crops as feed for ruminant and monogastric (one stomach; mainly chickens and pigs) livestock. Manure is typically recycled in mixed systems as a fertilizer for crops.

Landless systems rely upon feed from outside the farm, representing the de-linking of crop and livestock production found more prevalently in Organisation for Economic Co-operation and Development(OECD) member countries. Synthetic fertilizers are more heavily relied upon for crop production and manure utilization becomes a challenge as well as a source for pollution. Industrialized countries use these operations to produce much of the global supplies of poultry and pork. Scientists estimate that 75% of the growth in livestock production between 2003 and 2030 will be in confined animal feeding operations, sometimes called factory farming. Much of this growth is happening in developing countries in Asia, with much smaller amounts of growth in Africa. Some of the practices used in commercial livestock production, including the usage of growth hormones, are controversial.

Production Practices

Road leading across the farm allows machinery access to the farm for production practices

Farming is the practice of agriculture by specialized labor in an area primarily devoted to agricultural processes, in service of a dislocated population usually in a city.

Tillage is the practice of plowing soil to prepare for planting or for nutrient incorporation or for pest control. Tillage varies in intensity from conventional to no-till. It may improve productivity by warming the soil, incorporating fertilizer and controlling weeds, but also renders soil more prone to erosion, triggers the decomposition of organic matter releasing CO_2, and reduces the abundance and diversity of soil organisms.

Pest control includes the management of weeds, insects, mites, and diseases. Chemical (pesticides), biological (biocontrol), mechanical (tillage), and cultural practices are used. Cultural practices include crop rotation, culling, cover crops, intercropping, composting, avoidance, and resistance. Integrated pest management attempts to use all of these methods to keep pest populations below the number which would cause economic loss, and recommends pesticides as a last resort.

Nutrient management includes both the source of nutrient inputs for crop and livestock production, and the method of utilization of manure produced by livestock. Nutrient inputs can be chemical inorganic fertilizers, manure, green manure, compost and mined minerals. Crop nutrient use may also be managed using cultural techniques such as crop rotation or a fallow period. Manure is used either by holding livestock where the feed crop is growing, such as in managed intensive rotational grazing, or by spreading either dry or liquid formulations of manure on cropland or pastures.

Water management is needed where rainfall is insufficient or variable, which occurs to some degree in most regions of the world. Some farmers use irrigation to supplement rainfall. In other areas such as the Great Plains in the U.S. and Canada, farmers use a fallow year to conserve soil moisture to use for growing a crop in the following year. Agriculture represents 70% of freshwater use worldwide.

According to a report by the International Food Policy Research Institute, agricultural technologies will have the greatest impact on food production if adopted in combination with each other; using a model that assessed how eleven technologies could impact agricultural productivity, food security and trade by 2050, the International Food Policy Research Institute found that the number of people at risk from hunger could be reduced by as much as 40% and food prices could be reduced by almost half.

"Payment for ecosystem services (PES) can further incentivise efforts to green the agriculture sector. This is an approach that verifies values and rewards the benefits of ecosystem services provided by green agricultural practices." "Innovative PES measures could include reforestation payments made by cities to upstream communities in rural areas of shared watersheds for improved quantities and quality of fresh water for municipal users. Ecoservice payments by farmers to upstream forest stewards for properly managing the flow of soil nutrients, and methods to monetise the carbon sequestration and emission reduction credit benefits of green agriculture practices in order to compensate farmers for their efforts to restore and build SOM and employ other practices."

Crop Alteration and Biotechnology

Tractor and chaser bin

Crop alteration has been practiced by humankind for thousands of years, since the beginning of civilization. Altering crops through breeding practices changes the genetic make-up of a plant to develop crops with more beneficial characteristics for humans, for example, larger fruits or seeds, drought-tolerance, or resistance to pests. Significant advances in plant breeding ensued after the work of geneticist Gregor Mendel. His work on dominant and recessive alleles, although initially largely ignored for almost 50 years, gave plant breeders a better understanding of genetics and breeding techniques. Crop breeding includes techniques such as plant selection with desirable traits, self-pollination and cross-pollination, and molecular techniques that genetically modify the organism.

Domestication of plants has, over the centuries increased yield, improved disease resistance and drought tolerance, eased harvest and improved the taste and nutritional value of crop plants. Careful selection and breeding have had enormous effects on the characteristics of crop plants. Plant selection and breeding in the 1920s and 1930s improved pasture (grasses and clover) in New Zealand. Extensive X-ray and ultraviolet induced mutagenesis efforts (i.e. primitive genetic engineering) during the 1950s produced the modern commercial varieties of grains such as wheat, corn (maize) and barley.

The Green Revolution popularized the use of conventional hybridization to sharply increase yield by creating "high-yielding varieties". For example, average yields of corn (maize) in the USA have increased from around 2.5 tons per hectare (t/ha) (40 bushels per acre) in 1900 to about 9.4 t/ha (150 bushels per acre) in 2001. Similarly, worldwide average wheat yields have increased from less than 1 t/ha in 1900 to more than 2.5 t/ha in 1990. South American average wheat yields are around 2 t/ha, African under 1 t/ha, and Egypt and Arabia up to 3.5 to 4 t/ha with irrigation. In contrast, the average wheat yield in countries such as France is over 8 t/ha. Variations in yields are due mainly to variation in climate, genetics, and the level of intensive farming techniques (use of fertilizers, chemical pest control, growth control to avoid lodging).

Genetic Engineering

Genetically modified organisms (GMO) are organisms whose genetic material has been altered by genetic engineering techniques generally known as recombinant DNA technology. Genetic engineering has expanded the genes available to breeders to utilize in creating desired germlines for new crops. Increased durability, nutritional content, insect and virus resistance and herbicide tolerance are a few of the attributes bred into crops through genetic engineering. For some, GMO crops cause food safety and food labeling concerns. Numerous countries have placed restrictions on the production, import or use of GMO foods and crops, which have been put in place due to concerns over potential health issues, declining agricultural diversity and contamination of non-GMO crops. Currently a global treaty, the Biosafety Protocol, regulates the trade of GMOs. There is ongoing discussion regarding the labeling of foods made from GMOs, and while the EU currently requires all GMO foods to be labeled, the US does not.

Herbicide-resistant seed has a gene implanted into its genome that allows the plants to tolerate exposure to herbicides, including glyphosates. These seeds allow the farmer to grow a crop that can be sprayed with herbicides to control weeds without harming the resistant crop. Herbicide-tolerant crops are used by farmers worldwide. With the increasing use of herbicide-tolerant crops, comes an increase in the use of glyphosate-based herbicide sprays. In some areas glyphosate resistant weeds have developed, causing farmers to switch to other herbicides. Some studies also link widespread glyphosate usage to iron deficiencies in some crops, which is both a crop production and a nutritional quality concern, with potential economic and health implications.

Other GMO crops used by growers include insect-resistant crops, which have a gene from the soil bacterium *Bacillus thuringiensis* (Bt), which produces a toxin specific to insects. These crops protect plants from damage by insects. Some believe that similar or better pest-resistance traits can be acquired through traditional breeding practices, and resistance to various pests can be gained through hybridization or cross-pollination with wild species. In some cases, wild species are the primary source of resistance traits; some tomato cultivars that have gained resistance to at least 19 diseases did so through crossing with wild populations of tomatoes.

Environmental Impact

Water pollution in a rural stream due to runoff from farming activity in New Zealand

Agriculture, as implemented through the method of farming, imposes external costs upon society through pesticides, nutrient runoff, excessive water usage, loss of natural environment and assorted other problems. A 2000 assessment of agriculture in the UK determined total external costs for 1996 of £2,343 million, or £208 per hectare. A 2005 analysis of these costs in the USA concluded that cropland imposes approximately $5 to 16 billion ($30 to $96 per hectare), while livestock production imposes $714 million. Both studies, which focused solely on the fiscal impacts, concluded that more should be done to internalize external costs. Neither included subsidies in their analysis, but they noted that subsidies also influence the cost of agriculture to society. In 2010, the International Resource Panel of the United Nations Environment Programme published a report assessing the environmental impacts of consumption and production. The study found that agriculture and food consumption are two of the most important drivers of environmental pressures, particularly habitat change, climate change, water use and toxic emissions. The 2011 UNEP Green

Economy report states that "[a]gricultural operations, excluding land use changes, produce approximately 13 per cent of anthropogenic global GHG emissions. This includes GHGs emitted by the use of inorganic fertilisers agro-chemical pesticides and herbicides; (GHG emissions resulting from production of these inputs are included in industrial emissions); and fossil fuel-energy inputs. "On average we find that the total amount of fresh residues from agricultural and forestry production for second-generation biofuel production amounts to 3.8 billion tonnes per year between 2011 and 2050 (with an average annual growth rate of 11 per cent throughout the period analysed, accounting for higher growth during early years, 48 per cent for 2011–2020 and an average 2 per cent annual expansion after 2020)."

Livestock Issues

A senior UN official and co-author of a UN report detailing this problem, Henning Steinfeld, said "Livestock are one of the most significant contributors to today's most serious environmental problems". Livestock production occupies 70% of all land used for agriculture, or 30% of the land surface of the planet. It is one of the largest sources of greenhouse gases, responsible for 18% of the world's greenhouse gas emissions as measured in CO_2 equivalents. By comparison, all transportation emits 13.5% of the CO_2. It produces 65% of human-related nitrous oxide (which has 296 times the global warming potential of CO_2.) and 37% of all human-induced methane (which is 23 times as warming as CO_2.) It also generates 64% of the ammonia emission. Livestock expansion is cited as a key factor driving deforestation; in the Amazon basin 70% of previously forested area is now occupied by pastures and the remainder used for feedcrops. Through deforestation and land degradation, livestock is also driving reductions in biodiversity. Furthermore, the UNEP states that "methane emissions from global livestock are projected to increase by 60 per cent by 2030 under current practices and consumption patterns."

Land and Water Issues

Land transformation, the use of land to yield goods and services, is the most substantial way humans alter the Earth's ecosystems, and is considered the driving force in the loss of biodiversity. Estimates of the amount of land transformed by humans vary from 39 to 50%. Land degradation, the long-term decline in ecosystem function and productivity, is estimated to be occurring on 24% of land worldwide, with cropland overrepresented. The UN-FAO report cites land management as the driving factor behind degradation and reports that 1.5 billion people rely upon the degrading land. Degradation can be deforestation, desertification, soil erosion, mineral depletion, or chemical degradation (acidification and salinization).

Eutrophication, excessive nutrients in aquatic ecosystems resulting in algal blooms and anoxia, leads to fish kills, loss of biodiversity, and renders water unfit for drinking and other industrial uses. Excessive fertilization and manure application to cropland, as well as high livestock stocking densities cause nutrient (mainly nitrogen and phosphorus) runoff and leaching from agricultural land. These nutrients are major nonpoint pollutants contributing to eutrophication of aquatic ecosystems.

Agriculture accounts for 70 percent of withdrawals of freshwater resources. Agriculture is a major draw on water from aquifers, and currently draws from those underground water sources at an unsustainable rate. It is long known that aquifers in areas as diverse as northern China, the Upper Ganges and the western US are being depleted, and new research extends these problems to aquifers in Iran, Mexico and Saudi Arabia. Increasing pressure is being placed on water resources by industry and urban areas, meaning that water scarcity is increasing and agriculture is facing the challenge of producing more food for the world's growing population with reduced water resources. Agricultural water usage can also cause major environmental problems, including the destruction of natural wetlands, the spread of water-borne diseases, and land degradation through salinization and waterlogging, when irrigation is performed incorrectly.

Pesticides

Pesticide use has increased since 1950 to 2.5 million short tons annually worldwide, yet crop loss from pests has remained relatively constant. The World Health Organization estimated in 1992 that 3 million pesticide poisonings occur annually, causing 220,000 deaths. Pesticides select for pesticide resistance in the pest population, leading to a condition termed the "pesticide treadmill" in which pest resistance warrants the development of a new pesticide.

An alternative argument is that the way to "save the environment" and prevent famine is by using pesticides and intensive high yield farming, a view exemplified by a quote heading the Center for Global Food Issues website: 'Growing more per acre leaves more land for nature'. However, critics argue that a trade-off between the environment and a need for food is not inevitable, and that pesticides simply replace good agronomic practices such as crop rotation. The UNEP introduces the Push–pull agricultural pest management technique which involves intercropping that uses plant aromas to repel or push away pests while pulling in or attracting the right insects. "The implementation of push-pull in eastern Africa has significantly increased maize yields and the combined cultivation of N-fixing forage crops has enriched the soil and has also provided farmers with feed for livestock. With increased livestock operations, the farmers are able to produce meat, milk and other dairy products and they use the manure as organic fertiliser that returns nutrients to the fields."

Climate Change

Climate change has the potential to affect agriculture through changes in temperature, rainfall (timing and quantity), CO_2, solar radiation and the interaction of these elements. Extreme events, such as droughts and floods, are forecast to increase as climate change takes hold. Agriculture is among sectors most vulnerable to the impacts of climate change; water supply for example, will be critical to sustain agricultural production and provide the increase in food output required to sustain the world's growing population. Fluctuations in the flow of rivers are likely to increase in the twenty-first century. Based on the experience of countries in the Nile river basin (Ethiopia, Kenya and Sudan) and other developing countries, depletion of water resources during seasons crucial for agriculture can lead to a decline in yield by up to 50%. Transformational approaches will be needed to manage natural resources in the future. For example, policies, practices and tools promoting climate-smart agriculture will be important, as will better use of scientific information on climate for assessing risks and vulnerability. Planners and policy-makers will need to help create suitable policies that encourage funding for such agricultural transformation.

Agriculture in its many forms can both mitigate or worsen global warming. Some of the increase in CO_2 in the atmosphere comes from the decomposition of organic matter in the soil, and much of the methane emitted into the atmosphere is caused by the decomposition of organic matter in wet soils such as rice paddy fields, as well as the normal digestive activities of farm animals. Further, wet or anaerobic soils also lose nitrogen through denitrification, releasing the greenhouse gases nitric oxide and nitrous oxide. Changes in management can reduce the release of these greenhouse gases, and soil can further be used to sequester some of the CO_2 in the atmosphere. Informed by the UNEP, "[a]griculture also produces about 58 per cent of global nitrous oxide emissions and about 47 per cent of global methane emissions. Cattle and rice farms release methane, fertilized fields release nitrous oxide, and the cutting down of rainforests to grow crops or raise livestock releases carbon dioxide. Both of these gases have a far greater global warming potential per tonne than CO2 (298 times and 25 times respectively)."

There are several factors within the field of agriculture that contribute to the large amount of CO2 emissions. The diversity of the sources ranges from the production of farming tools to the transport of harvested produce. Approximately 8% of the national carbon footprint is due to agricultural sources. Of that, 75% is of the carbon emissions released from the production of crop assisting chemicals. Factories producing insecticides, herbicides, fungicides, and fertilizers are a major culprit of the greenhouse gas. Productivity on the farm itself and the use of machinery is another source of the carbon emission. Almost all the industrial machines used in modern farming are powered by fossil fuels. These instruments are burning fossil fuels from the beginning of the process to the end. Tractors are the root of this source. The tractor is going to burn fuel and release CO2 just to run. The amount of emissions from the machinery increase with the attachment of different units and need for more power. During the soil preparation stage tillers and plows will be used to disrupt the soil. During growth watering pumps and sprayers are used to keep the crops hydrated. And when the crops are ready for picking a forage or combine harvester is used. These types of machinery all require additional energy which leads to increased carbon dioxide emissions from the basic tractors. The final major contribution to CO_2 emissions in agriculture is in the final transport of produce. Local farming suffered a decline over the past century due to large amounts of farm subsidies. The majority of crops are shipped hundreds of miles to various processing plants before ending up in the grocery store. These shipments are made using fossil fuel burning modes of transportation. Inevitably these transport adds to carbon dioxide emissions.

Sustainability

Some major organizations are hailing farming within agroecosystems as the way forward for mainstream agriculture. Current farming methods have resulted in over-stretched water resources, high levels of erosion and reduced soil fertility. According to a report by the International Water Management Institute and UNEP, there is not enough water to continue farming using current practices; therefore how critical water, land, and ecosystem resources are used to boost crop yields must be reconsidered. The report suggested assigning value to ecosystems, recognizing environmental and livelihood tradeoffs, and balancing the rights of a variety of users and interests. Inequities that result when such measures are adopted would need to be addressed, such as the reallocation of water from poor to rich, the clearing of land to make way for more productive farmland, or the preservation of a wetland system that limits fishing rights.

Technological advancements help provide farmers with tools and resources to make farming more sustainable. New technologies have given rise to innovations like conservation tillage, a farming process which helps prevent land loss to erosion, water pollution and enhances carbon sequestration.

According to a report by the International Food Policy Research Institute (IFPRI), agricultural technologies will have the greatest impact on food production if adopted in combination with each other; using a model that assessed how eleven technologies could impact agricultural productivity, food security and trade by 2050, IFPRI found that the number of people at risk from hunger could be reduced by as much as 40% and food prices could be reduced by almost half.

Agricultural Economics

Agricultural economics refers to economics as it relates to the "production, distribution and consumption of [agricultural] goods and services". Combining agricultural production with general theories of marketing and business as a discipline of study began in the late 1800s, and grew significantly through the 20th century. Although the study of agricultural economics is relatively recent, major trends in agriculture have significantly affected national and international economies throughout history, ranging from tenant farmers and sharecropping in the post-American Civil War Southern United States to the European feudal system of manorialism. In the United States, and elsewhere, food costs attributed to food processing, distribution, and agricultural marketing, sometimes referred to as the value chain, have risen while the costs attributed to farming have declined. This is related to the greater efficiency of farming, combined with the increased level of value addition (e.g. more highly processed products) provided by the supply chain. Market concentration has increased in the sector as well, and although the total effect of the increased market concentration is likely increased efficiency, the changes redistribute economic surplus from producers (farmers) and consumers, and may have negative implications for rural communities.

National government policies can significantly change the economic marketplace for agricultural products, in the form of taxation, subsidies, tariffs and other measures. Since at least the 1960s, a combination of import/export restrictions, exchange rate policies and subsidies have affected farmers in both the developing and developed world. In the 1980s, it was clear that non-subsidized farmers in developing countries were experiencing adverse effects from national policies that created artificially low global prices for farm products. Between the mid-1980s and the early 2000s, several international agreements were put into place that limited agricultural tariffs, subsidies and other trade restrictions.

However, as of 2009, there was still a significant amount of policy-driven distortion in global agricultural product prices. The three agricultural products with the greatest amount of trade distortion were sugar, milk and rice, mainly due to taxation. Among the oilseeds, sesame had the greatest amount of taxation, but overall, feed grains and oilseeds had much lower levels of taxation than livestock products. Since the 1980s, policy-driven distortions have seen a greater decrease among livestock products than crops during the worldwide reforms in agricultural policy. Despite this progress, certain crops, such as cotton, still see subsidies in developed countries artificially deflating global prices, causing hardship in developing countries with non-subsidized farmers. Unprocessed commodities (i.e. corn, soybeans, cows) are generally graded to indicate quality. The quality affects the price the producer receives. Commodities are generally reported by production quantities, such as volume, number or weight.

Agricultural Science

Agricultural science is a broad multidisciplinary field of biology that encompasses the parts of exact, natural, economic and social sciences that are used in the practice and understanding of agriculture. (Veterinary science, but not animal science, is often excluded from the definition.)

List of Countries by Agricultural Output

Energy and Agriculture

Since the 1940s, agricultural productivity has increased dramatically, due largely to the increased use of energy-intensive mechanization, fertilizers and pesticides. The vast majority of this energy input comes from fossil fuel sources. Between the 1960–65 measuring cycle and the cycle from 1986 to 1990, the Green Revolution transformed agriculture around the globe, with world grain production increasing significantly (between 70% and 390% for wheat and 60% to 150% for rice, depending on geographic area) as world population doubled. Modern agriculture's heavy reliance on petrochemicals and mechanization has raised concerns that oil shortages could increase costs and reduce agricultural output, causing food shortages.

Agriculture and food system share (%) of total energy consumption by three industrialized nations			
Country	**Year**	**Agriculture (direct & indirect)**	**Food system**
United Kingdom	2005	1.9	11
United States	1996	2.1	10
United States	2002	2.0	14
Sweden	2000	2.5	13

Modern or industrialized agriculture is dependent on fossil fuels in two fundamental ways: 1. direct consumption on the farm and 2. indirect consumption to manufacture inputs used on the farm. Direct consumption includes the use of lubricants and fuels to operate farm vehicles and machinery; and use of gasoline, liquid propane, and electricity to power dryers, pumps, lights, heaters, and coolers. American farms directly consumed about 1.2 exajoules (1.1 quadrillion BTU) in 2002, or just over 1% of the nation's total energy.

Indirect consumption is mainly oil and natural gas used to manufacture fertilizers and pesticides, which accounted for 0.6 exajoules (0.6 quadrillion BTU) in 2002. The natural gas and coal consumed by the production of nitrogen fertilizer can account for over half of the agricultural energy usage. China utilizes mostly coal in the production of nitrogen fertilizer, while most of Europe uses large amounts of natural gas and small amounts of coal. According to a 2010 report published by The Royal Society, agriculture is increasingly dependent on the direct and indirect input of fossil fuels. Overall, the fuels used in agriculture vary based on several factors, including crop, production system and location. The energy used to manufacture farm machinery is also a form of indirect agricultural energy consumption. Together, direct and indirect consumption by US farms accounts for about 2% of the nation's energy use. Direct and indirect energy consumption by U.S. farms peaked in 1979, and has gradually declined over the past 30 years. Food systems encompass not just agricultural production, but also off-farm processing, packaging, transporting, marketing, consumption, and disposal of food and food-related items. Agriculture accounts for less than one-fifth of food system energy use in the US.

Mitigation of Effects of Petroleum Shortages

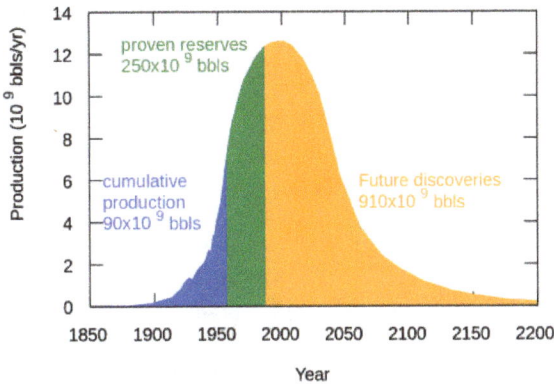

M. King Hubbert's prediction of world petroleum production rates. Modern agriculture is totally reliant on petroleum energy

In the event of a petroleum shortage, organic agriculture can be more attractive than conventional practices that use petroleum-based pesticides, herbicides, or fertilizers. Some studies using modern organic-farming methods have reported yields equal to or higher than those available from conventional farming. In the aftermath of the fall of the Soviet Union, with shortages of conventional petroleum-based inputs, Cuba made use of mostly organic practices, including biopesticides, plant-based pesticides and sustainable cropping practices, to feed its populace. However, organic farming may be more labor-intensive and would require a shift of the workforce from urban to rural areas. The reconditioning of soil to restore organic matter lost during the use of monoculture agriculture techniques is important to provide a reservoir of plant-available nutrients, to maintain texture, and to minimize erosion.

It has been suggested that rural communities might obtain fuel from the biochar and synfuel process, which uses agricultural *waste* to provide charcoal fertilizer, some fuel *and* food, instead of the normal food vs. fuel debate. As the synfuel would be used on-site, the process would be more efficient and might just provide enough fuel for a new organic-agriculture fusion.

It has been suggested that some transgenic plants may some day be developed which would allow for maintaining or increasing yields while requiring fewer fossil-fuel-de-rived inputs than conventional crops. The possibility of success of these programs is questioned by ecologists and economists concerned with unsustainable GMO prac-tices such as terminator seeds. While there has been some research on sustainability using GMO crops, at least one prominent multi-year attempt by Monsanto Company has been unsuccessful, though during the same period traditional breeding techniques yielded a more sustainable variety of the same crop.

Policy

United States farm subsidies in 2005

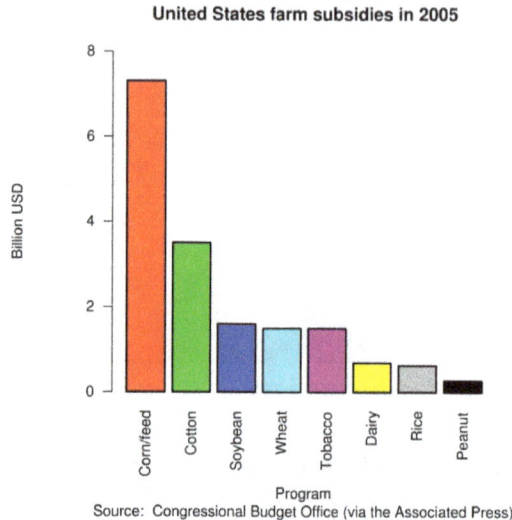

Source: Congressional Budget Office (via the Associated Press)

From a Congressional Budget Office report

Agricultural policy is the set of government decisions and actions relating to domestic agriculture and imports of foreign agricultural products. Governments usually implement agricultural policies with the goal of achieving a specific outcome in the domestic agricultural product markets. Some overarching themes include risk management and adjustment (including policies related to climate change, food safety and natural disasters), economic stability (including policies related to taxes), natural resources and environmental sustainability (especially water policy), research and development, and market access for domestic commodities (including relations with global organizations and agreements with other countries). Agricultural policy can also touch on food quality, ensuring that the food supply is of a consistent and known quality, food security, ensuring that the food supply meets the population's needs, and conservation. Policy programs can range from financial programs, such as subsidies, to encouraging producers to enroll in voluntary quality assurance programs.

There are many influences on the creation of agricultural policy, including consumers, agribusiness, trade lobbies and other groups. Agribusiness interests hold a large amount of influence over policy making, in the form of lobbying and campaign contributions. Political action groups, including those interested in environmental issues and labor unions, also provide influence, as do lobbying organizations representing individual agricultural commodities. The Food and Agriculture Organization of the United Nations (FAO) leads international efforts to defeat hunger and provides a forum for the negotiation of global agricultural regulations and agreements. Dr. Samuel Jutzi, director of FAO's animal production and health division, states that lobbying by large corporations has stopped reforms that would improve human health and the environment. For example, proposals in 2010 for a voluntary code of conduct for the livestock industry that would have provided incentives for improving standards for health, and environmental regulations, such as the number of animals an area of land can support without long-term damage, were successfully defeated due to large food company pressure.

Forestry

Forestry work in Austria

Forestry is the science and craft of creating, managing, using, conserving, and repairing forests and associated resources to meet desired goals, needs, and values for human and environment benefits. Forestry is practiced in plantations and natural stands. The science of forestry has elements that belong to the biological, physical, social, political and managerial sciences.

Modern forestry generally embraces a broad range of concerns, in what is known as multiple-use management, including the provision of timber, fuel wood, wildlife habitat, natural water quality management, recreation, landscape and community protection, employment, aesthetically appealing landscapes, biodiversity management, watershed management, erosion control, and preserving forests as 'sinks' for atmospheric carbon dioxide. A practitioner of forestry is known as a forester. Other terms are used a verderer and a silviculturalist being common ones. Silviculture is narrower than forestry, being concerned only with forest plants, but is often used synonymously with forestry.

Forest ecosystems have come to be seen as the most important component of the biosphere, and forestry has emerged as a vital applied science, craft, and technology.

Forestry is an important economic segment in various industrial countries. For example, in Germany, forests cover nearly a third of the land area, wood is the most important renewable resource, and forestry supports more than a million jobs and about billion in yearly turnover.

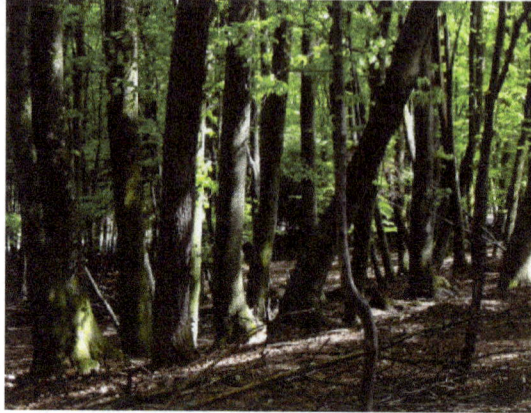

A deciduous beech forest in Slovenia

History

Background

The preindustrial age has been dubbed by Werner Sombart and others as the 'wooden age', as timber and firewood were the basic resources for energy, construction and housing. The development of modern forestry is closely connected with the rise of capitalism, economy as a science and varying notions of land use and property.

Roman Latifundiae, large agricultural estates, were quite successful in maintaining the large supply of wood that was necessary for the Roman Empire. Large deforestations came with respectively after the decline of the Romans. However already in the 5th century, monks in the then Byzantine Romagna on the Adriatic coast, were able to establish stone pine plantations to provide fuelwood and food. This was the beginning of the massive forest mentioned by Dante Alighieri in his 1308 poem Divine Comedy.

Similar sustainable formal forestry practices were developed by the Visigoths in the 7th century when, faced with the ever increasing shortage of wood, they instituted a code concerned with the preservation of oak and pine forests. The use and management of many forest resources has a long history in China as well, dating back to the Han Dynasty and taking place under the landowning gentry. A similar approach was used in Japan. It was also later written about by the Ming Dynasty Chinese scholar Xu Guangqi (1562–1633).

In Europe, land usage rights in medieval and early modern times allowed different users to access forests and pastures. Plant litter and resin extraction were important, as pitch (resin) was essential for the caulking of ships, falking and hunting rights, firewood and building, timber gathering in wood pastures, and for grazing animals in forests. The notion of "commons" (German "Allmende") refers to the underlying traditional legal term of common land. The idea of enclosed private property came about during modern times. However, most hunting rights were retained by members of the nobility which preserved the right of the nobility to access and use common land for recreation, like fox hunting.

Early Modern Forestry Development

Timber harvesting, as here in Finland, is a common component of forestry

Hans Carl von Carlowitz

Systematic management of forests for a sustainable yield of timber is said to have begun in the German states in the 14th century, e.g. in Nuremberg, and in 16th-century Japan. Typically, a forest was divided into specific sections and mapped; the harvest of timber was planned with an eye to regeneration. As timber rafting allowed for connecting large continental forests, as in south western Germany, via Main, Neckar, Danube and Rhine with the coastal cities and states, early modern forestry and remote trading were closely connected. Large firs in the black forest were called „Holländer", as they were traded to the Dutch ship yards. Large timber rafts on the Rhine were 200 to 400m in length, 40m in width and consisted of several thousand logs. The crew consisted of 400 to 500 men, including shelter, bakeries, ovens and livestock stables. Timber rafting infrastructure allowed for large interconnected networks all over continental Europe and is still of importance in Finland.

Starting with the sixteenth century, enhanced world maritime trade, a boom in housing construction in Europe and the success and further Berggeschrey (rushes) of the mining industry increased timber consumption sharply. The notion of 'Nachhaltigkeit', sustainability in forestry, is closely connected to the work of Hans Carl von Carlowitz (1645–1714), a mining administrator in Saxony. His book *Sylvicultura oeconomica, oder haußwirthliche Nachricht und Naturmäßige Anweisung zur wilden Baum-Zucht* (1713) was the first comprehensive treatise about sustainable yield forestry. In the UK, and to an extend in continental Europe, the enclosure movement and the clearances favored strictly enclosed private property. The Agrarian reformers, early economic writers and scientists tried to get rid of the traditional commons. At the time, an alleged tragedy of the commons together with fears of a Holznot, an imminent wood shortage played a watershed role in the controversies about cooperative land use patterns.

The practice of establishing tree plantations in the British Isles was promoted by John Evelyn, though it had already acquired some popularity. Louis XIV's minister Jean-Baptiste Colbert's oak Forest of Tronçais, planted for the future use of the French Navy, matured as expected in the mid-19th century: "Colbert had thought of everything except the steamship," Fernand Braudel observed. In parallel, schools of forestry were established beginning in the late 18th century in Hesse, Russia, Austria-Hungary, Sweden, France and elsewhere in Europe.

Forest Conservation and Early Globalization

During the late 19th and early 20th centuries, forest preservation programs were established in British India, the United States, and Europe. Many foresters were either from continental Europe (like Sir Dietrich Brandis), or educated there (like Gifford Pinchot). Sir Dietrich Brandis is considered the father of tropical forestry, European concepts and practices had to be adapted in tropical and semi arid climate zones. The development of plantation forestry was one of the (controversial) answers to the specific challenges in the tropical colonies. The enactment and evolution of forest laws and binding regulations occurred in most Western nations in the 20th century in response to growing conservation concerns and the increasing technological capacity of logging companies. Tropical forestry is a separate branch of forestry which deals mainly with equatorial forests that yield woods such as teak and mahogany.

Mechanization

Forestry mechanization was always in close connection to metal working and the development of mechanical tools to cut and transport timber to its destination. Rafting belongs to the earliest means of transport. Steel saws came up in the 15th century. The 19th century widely increased the availability of steel for whipsaws and introduced Forest railways and railways in general for transport and as forestry customer. Further human induced changes, however, came since World War II, respectively in line with the '1950s-syndrome'. The first portable chainsaw was invented in 1918 in Canada, but large impact of mechanization in forestry started after World War II. Forestry Harvesters are among the most recent developments. Although drones, planes, laser scanning, satellites and robots also play a part in forestry.

Forestry Today

A modern sawmill

Today a strong body of research exists regarding the management of forest ecosystems and genetic improvement of tree species and varieties. Forestry also includes the development of better methods for the planting, protecting, thinning, controlled burning, felling, extracting, and processing of timber. One of the applications of modern forestry is reforestation, in which trees are planted and tended in a given area.

Trees provide numerous environmental, social and economic benefits for people. In many regions the forest industry is of major ecological, economic, and social importance. Third-party certification systems that provide independent verification of sound forest stewardship and sustainable forestry have become commonplace in many areas since the 1990s. These certification systems were developed as a response to criticism of some forestry practices, particularly deforestation in less developed regions along with concerns over resource management in the developed world. Some certification systems are criticized for primarily acting as marketing tools and lacking in their claimed independence.

In topographically severe forested terrain, proper forestry is important for the prevention or minimization of serious soil erosion or even landslides. In areas with a high potential for landslides, forests can stabilize soils and prevent property damage or loss, human injury, or loss of life.

Public perception of forest management has become controversial, with growing public concern over perceived mismanagement of the forest and increasing demands that forest land be managed for uses other than pure timber production, for example, indigenous rights, recreation, watershed management, and preservation of wilderness, waterways and wildlife habitat. Sharp disagreements over the role of forest fires, logging, motorized recreation and other issues drives debate while the public demand for wood products continues to increase.

Foresters

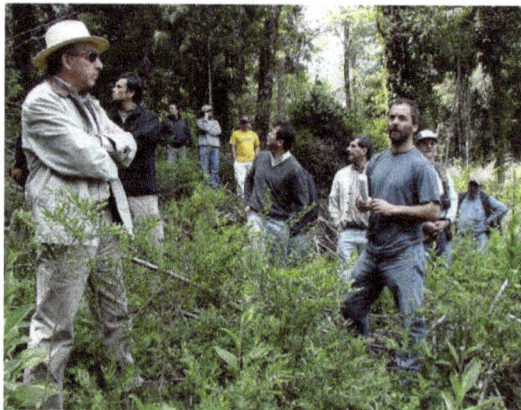

Foresters of UACh in the Valdivian forests of San Pablo de Tregua, Chile

Foresters work for the timber industry, government agencies, conservation groups, local authorities, urban parks boards, citizens' associations, and private landowners. The forestry profession includes a wide diversity of jobs, with educational requirements ranging from college bachelor's degrees to PhDs for highly specialized work. Industrial foresters plan forest regeneration starting with careful harvesting. Urban foresters manage trees in urban green spaces. Foresters work in tree nurseries growing seedlings for woodland creation or regeneration projects. Foresters improve tree genetics. Forest engineers develop new building systems. Professional foresters measure and model the growth of forests with tools like geographic information systems. Foresters may combat insect infestation, disease, forest and grassland wildfire, but increasingly allow these natural aspects of forest ecosystems to run their course when the likelihood of epidemics or risk of life or property are low. Increasingly, foresters participate in wildlife conservation planning and watershed protection. Foresters have been mainly concerned with timber management, especially reforestation, maintaining forests at prime conditions, and fire control.

Forestry Plans

Foresters develop and implement forest management plans relying on mapped resource inventories showing an area's topographical features as well as its distribution of trees (by species) and other plant cover. Plans also include landowner objectives, roads, culverts, proximity to human habitation, water features and hydrological conditions, and soils information. Forest management plans typically include recommended silvicultural treatments and a timetable for their implementation. Application of digital maps in Geographic Informations systems (GIS) that extracts and integrates different information about forest terrains, soil type and tree covers, etc. using, e.g. laser scanning, enhances forest management plans in modern systems.

Forest management plans include recommendations to achieve the landowner's objectives and desired future condition for the property subject to ecological, financial, logistical (e.g. access to resources), and other constraints. On some properties, plans focus on producing quality wood products for processing or sale. Hence, tree species, quantity, and form, all central to the value of harvested products quality and quantity, tend to be important components of silvicultural plans.

Good management plans include consideration of future conditions of the stand after any recommended harvests treatments, including future treatments (particularly in intermediate stand treatments), and plans for natural or artificial regeneration after final harvests.

The objectives of landowners and leaseholders influence plans for harvest and subsequent site treatment. In Britain, plans featuring "good forestry practice" must always consider the needs of other stakeholders such as nearby communities or rural residents living within or adjacent to woodland areas. Foresters consider tree felling and environmental legislation when developing plans. Plans instruct the sustainable harvesting and replacement of trees. They indicate whether road building or other forest engineering operations are required.

Agriculture and forest leaders are also trying to understand how the climate change legislation will affect what they do. The information gathered will provide the data that will determine the role of agriculture and forestry in a new climate change regulatory system.

Forestry as a Science

Over the past centuries, forestry was regarded as a separate science. With the rise of ecology and environmental science, there has been a reordering in the applied sciences. In line with this view, forestry is one of three primary land-use sciences. The other two are agriculture and agroforestry. Under these headings, the fundamentals behind the management of natural forests comes by way of natural ecology. Forests or tree plantations, those whose primary purpose is the extraction of forest products, are planned and managed utilizing a mix of ecological and agroecological principles.

Education

History of Forestry Education

The first dedicated forestry school was established by Georg Ludwig Hartig at Hungen in the Wetterau, Hesse, in 1787, though forestry had been taught earlier in central Europe, including at the University of Giessen, in Hesse-Darmstadt.

In Spain, the first forestry school was the Forest Engineering School of Madrid (Escuela Técnica Superior de Ingenieros de Montes), founded in 1844.

The first in North America, the Biltmore Forest School was established near Asheville, North Carolina, by Carl A. Schenck on September 1, 1898, on the grounds of George W. Vanderbilt's Biltmore Estate. Another early school was the New York State College of Forestry, established at Cornell University just a few weeks later, in September 1898. Early 19th century North American foresters went to Germany to study forestry. Some early German foresters also emigrated to North America.

In South America the first forestry school was established in Brazil, in Viçosa, Minas Gerais, in 1962, and moved the next year to become a faculty at the Federal University of Paraná, in Curitiba.

Forestry Education Today

Today, forestry education typically includes training in general biology, botany, genetics, soil science, climatology, hydrology, economics and forest management. Education in the basics of sociology and political science is often considered an advantage.

Prescribed burning is used by foresters to reduce fuel loads

In India, forestry education is imparted in the agricultural universities and in Forest Research Institutes (deemed universities). Four year degree programmes are conducted in these universities at the undergraduate level. Masters and Doctorate degrees are also available in these universities.

In the United States, postsecondary forestry education leading to a Bachelor's degree or Master's degree is accredited by the Society of American Foresters.

In Canada the Canadian Institute of Forestry awards silver rings to graduates from accredited university BSc programs, as well as college and technical programs.

In many European countries, training in forestry is made in accordance with requirements of the Bologna Process and the European Higher Education Area.

The International Union of Forest Research Organizations is the only international organization that coordinates forest science efforts worldwide.

Agroforestry

Parkland in Burkina Faso: sorghum grown under *Faidherbia albida* and *Borassus akeassii* near Banfora

Agroforestry or agro-sylviculture is a land use management system in which trees or shrubs are grown around or among crops or pastureland. It combines shrubs and trees in agricultural and forestry technologies to create more diverse, productive, profitable, healthy, ecologically sound, and sustainable land-use systems.

As a Science

The theoretical base for agroforestry comes from ecology, via agroecology. From this perspective, agroforestry is one of the three principal land-use sciences. The other two are agriculture and forestry.

Agroforestry has a lot in common with intercropping. Both have two or more plant species (such as nitrogen-fixing plants) in close interaction, both provide multiple outputs, as a consequence, higher overall yields and, because a single application or input is shared, costs are reduced. Beyond these, there are gains specific to agroforestry.

Benefits

Agroforestry systems can be advantageous over conventional agricultural, and forest production methods. They can offer increased productivity, economic benefits, and more diversity in the ecological goods and services provided .(An example of this was seen in trying to conserve Milicia excelsa.)

Biodiversity in agroforestry systems is typically higher than in conventional agricultural systems. With two or more interacting plant species in a given land area, it creates a more complex habitat that can support a wider variety of birds, insects, and other animals. Depending upon the application, impacts of agroforestry can include:

- Reducing poverty through increased production of wood and other tree products for home consumption and sale

- Contributing to food security by restoring the soil fertility for food crops

- Cleaner water through reduced nutrient and soil runoff

- Countering global warming and the risk of hunger by increasing the number of drought-resistant trees and the subsequent production of fruits, nuts and edible oils

- Reducing deforestation and pressure on woodlands by providing farm-grown fuelwood

- Reducing or eliminating the need for toxic chemicals (insecticides, herbicides, etc.)

- Through more diverse farm outputs, improved human nutrition

- In situations where people have limited access to mainstream medicines, providing growing space for medicinal plants

- Increased crop stability

- Multifunctional site use i.e. crop production and animal grazing.

- Typically more drought resistant.

- Stabilises depleted soils from erosion

- Bioremediation

Agroforestry practices may also realize a number of other associated environmental goals, such as:

- Carbon sequestration

- Odour, dust, and noise reduction

- Green space and visual aesthetics

- Enhancement or maintenance of wildlife habitat

Adaptation to Climate Change

There is some evidence that, especially in recent years, poor smallholder farmers are turning to agroforestry as a mean to adapt to the impacts of climate change. A study from the CGIAR research program on Climate Change, Agriculture and Food Security (CCAFS) found from a survey of over 700 households in East Africa that at least 50% of those households had begun planting trees on their farms in a change from their practices 10 years ago. The trees ameliorate the effects of climate change by helping to stabilize erosion, improving water and soil quality and providing yields of fruit, tea, coffee, oil, fodder and medicinal products in addition to their usual harvest. Agroforestry was one of the most widely adopted adaptation strategies in the study, along with the use of improved crop varieties and intercropping.

Applications

Agroforestry represents a wide diversity in application and in practice. One listing includes over 50 distinct uses. The 50 or so applications can be roughly classified under a few broad headings. There are visual similarities between practices in different categories. This is expected as categorization is based around the problems addressed (countering winds, high rainfall, harmful insects, etc.) and the overall economic constraints and objectives (labor and other inputs costs, yield requirements, etc.). The categories include :

- Parklands

- Shade systems

- Crop-over-tree systems

- Alley cropping

- Strip cropping

- Fauna-based systems

- Boundary systems

- Taungyas

- Physical support systems

- Agroforests

- Wind break and shelterbelt.

Parkland

Parklands are visually defined by the presence of trees widely scattered over a large agricultural plot or pasture. The trees are usually of a single species with clear regional favorites. Among the beaks and benefits, the trees offer shade to grazing animals, protect crops against strong wind bursts, provide tree prunings for firewood, and are a roost for insect or rodent-eating birds.

There are other gains. Research with *Faidherbia albida* in Zambia showed that mature trees can sustain maize yields of 4.1 tonnes per hectare compared to 1.3 tonnes per hectare without these trees. Unlike other trees, Faidherbia sheds its nitrogen-rich leaves during the rainy crop growing season so it does not compete with the crop for light, nutrients and water. The leaves then regrow during the dry season and provide land cover and shade for crops.

Shade Systems

With shade applications, crops are purposely raised under tree canopies and within the resulting shady environment. For most uses, the understory crops are shade tolerant or the overstory trees have fairly open canopies. A conspicuous example is shade-grown coffee. This practice reduces weeding costs and improves the quality and taste of the coffee. Just because plants are grown under shade does not necessarily translate into lost or reduced yields. This is because the efficiency of photosynthesis drops off with increasing light intensity, and the rate of photosynthesis hardly increases once the light intensity is over about one tenth that of direct overhead sun. This means that plants under trees can still grow well even though they get less light. By having more than one level of vegetation, it is possible to get more photosynthesis, and overall yields, than with a single canopy layer.

Crop-over-tree Systems

Not commonly encountered, crop-over-tree systems employ woody perennials in the role of a cover crop. For this, small shrubs or trees pruned to near ground level are utilized. The purpose, as with any cover crop, is to increase in-soil nutrients and/or to reduce soil erosion.

Alley Cropping

Alley cropping corn fields between rows of walnut trees.

With alley cropping, crop strips alternate with rows of closely spaced tree or hedge species. Normally, the trees are pruned before planting the crop. The cut leafy material is spread over the crop area to provide nutrients for the crop. In addition to nutrients, the hedges serve as windbreaks and eliminate soil erosion.

Alley cropping has been shown to be advantageous in Africa, particularly in relation to improving maize yields in the sub-Saharan region. Use here relies upon the nitrogen fixing tree species *Sesbania sesban, euphorbia tricalii, Tephrosia vogelii, Gliricidia sepium* and *Faidherbia albida*. In one example, a ten-year experiment in Malawi showed that, by using the fertilizer tree Gliricidia (*Gliricidia sepium*) on land on which no mineral fertilizer was applied, maize yields averaged 3.3 tonnes per hectare as compared to one tonne per hectare in plots without fertilizer trees nor mineral fertilizers.

Strip Cropping

Strip cropping is similar to alley cropping in that trees alternate with crops. The difference is that, with alley cropping, the trees are in single row. With strip cropping, the trees or shrubs are planted in wide strip. The purpose can be, as with alley cropping, to provide nutrients, in leaf form, to the crop. With strip cropping, the trees can have a purely productive role, providing fruits, nuts, etc. while, at the same time, protecting nearby crops from soil erosion and harmful winds.

Fauna-based Systems

There are situations where trees benefit fauna. The most common examples are the silvopasture where cattle, goats, or sheep browse on grasses grown under trees. In hot climates, the animals are less stressed and put on weight faster when grazing in a cooler, shaded environment. Other variations have these animals directly eating the leaves of trees or shrubs.

~ 1970   2004

Silvopasture over the years (Australia).

There are similar systems for other types of fauna. Deer and hogs gain when living and feeding in a forest ecosystem, especially when the tree forage suits their dietary needs. Another variation, aquaforestry, is where trees shade fish ponds. In many cases, the fish eat the leaves or fruit from the trees.

Boundary Systems

A riparian buffer bordering a river in Iowa.

There are a number of applications that fall under the heading of a boundary system. These include the living fences, the riparian buffer, and windbreaks.

- A living fence can be a thick hedge or fencing wire strung on living trees. In addition to restricting the movement of people and animals, living fences offer habitat to insect-eating birds and, in the case of a boundary hedge, slow soil erosion.

- Riparian buffers are strips of permanent vegetation located along or near active watercourses or in ditches where water runoff concentrates. The purpose is to keep nutrients and soil from contaminating surface water.

- Windbreaks reduce the velocity of the winds over and around crops. This increases yields through reduced drying of the crop and/or by preventing the crop from toppling in strong wind gusts.

Taungya

Taungya is a vastly used system originating in Burma. In the initial stages of an orchard or tree plantation, the trees are small and widely spaced. The free space between the newly planted trees can accommodate a seasonal crop. Instead of costly weeding, the underutilized area provides an additional output and income. More complex taungyas use the between-tree space for a series of crops. The crops become more shade resistant as the tree canopies grow and the amount of sunlight reaching the ground declines. If a plantation is thinned in the latter stages, this opens further the between-tree cropping opportunities.

Physical Support Systems

In the long history of agriculture, trellises are comparatively recent. Before this, grapes and other vine crops were raised atop pruned trees. Variations of the physical support theme depend upon the type of vine. The advantages come through greater in-field biodiversity. In many cases, the control of weeds, diseases, and insect pests are primary motives.

Agroforests

These are widely found in the humid tropics and are referenced by different names (forest gardening, forest farming, tropical home gardens and, where short-statured trees or shrubs dominate, shrub gardens). Through a complex, diverse mix of trees, shrubs, vines, and seasonal crops, these systems achieve the ecological dynamics of a forest ecosystem. Because of their internal ecology, they tend to be less susceptible to harmful insects, plant diseases, drought, and wind damage.

Historical Use

Agroforestry similar methods were historically utilized by Native Americans. California Indians would prescribe burn oak and other habitats to maintain a 'pyrodiversity collecting model'. This method allowed for greater health of trees and the habitat in general.

Challenges

Agroforestry is relevant to almost all environments and is a potential response to common problems around the globe, and agroforestry systems can be advantageous compared to conventional agriculture or forestry. Yet agroforestry is not very widespread, at least according to current but incomplete USDA surveys as of November, 2013.

As suggested by a survey of extension programs in the United States, some obstacles (ordered most critical to least critical) to agroforestry adoption include:

- Lack of developed markets for products

- Unfamiliarity with technologies

- Lack of awareness of successful agroforestry examples

- Competition between trees, crops, and animals

- Lack of financial assistance

- Lack of apparent profit potential

- Lack of demonstration sites

- Expense of additional management

- Lack of training or expertise

- Lack of knowledge about where to market products

- Lack of technical assistance

- Cannot afford adoption or start up costs, including costs of time

- Unfamiliarity with alternative marketing approaches (e.g. web)

- Unavailability of information about agroforestry

- Apparent inconvenience

- Lack of infrastructure (e.g. buildings, equipment)

- Lack of equipment

- Insufficient land

- Lack of seed/seedling sources

- Lack of scientific research

Some solutions to these obstacles have already been suggested although many depend on particular circumstances which vary from one location to the next.

Agronomy

Agronomy (agrós 'field' + nómos 'law') is the science and tech-nology of producing and using plants for food, fuel, fiber, and land reclamation. Agronomy has come to encompass work in the areas of plant genetics, plant physiology, meteorology, and soil science. It is the application of a combination of sciences like biology, chemistry, economics, ecology, earth science, and genetics. Agronomists of today are involved with many issues, including producing food, creating healthier food, managing the environmental impact of agriculture, and extracting energy from plants. Agronomists often specialise in areas such as crop rotation, irrigation and drainage, plant breeding, plant physiology, soil classification, soil fertility, weed control, and insect and pest control.

Plant Breeding

An agronomist field sampling a trial plot of flax.

This area of agronomy involves selective breeding of plants to produce the best crops under various conditions. Plant breeding has increased crop yields and has improved the nutritional value of numerous crops, including corn, soybeans, and wheat. It has also led to the development of new types of plants. For example, a hybrid grain called triticale was produced by crossbreeding rye and wheat. Triticale contains more usable protein than does either rye or wheat. Agronomy has also been instrumental in fruit and vegetable production research.

Biotechnology

Purdue University agronomy professor George Van Scoyoc explains the difference between forest and prairie soils to soldiers of the Indiana National Guard's Agribusiness Development Team at the Beck Agricultural Center in West Lafayette, Indiana

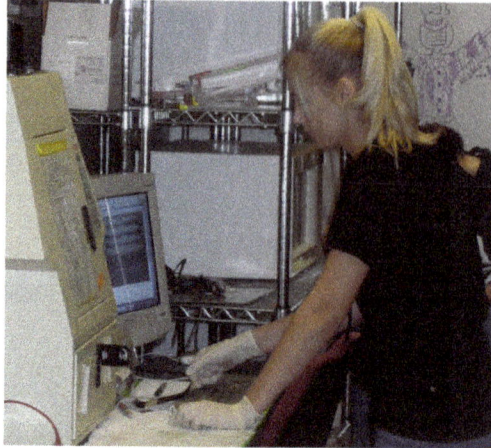

An agronomist mapping a plant genome

Agronomists use biotechnology to extend and expedite the development of desired characteristic. Biotechnology is often a lab activity requiring field testing of the new crop varieties that are developed.

In addition to increasing crop yields agronomic biotechnology is increasingly being applied for novel uses other than food. For example, oilseed is at present used mainly for margarine and other food oils, but it can be modified to produce fatty acids for detergents, substitute fuels and petrochemicals.

Soil Science

Agronomists study sustainable ways to make soils more productive and profitable. They classify soils and analyze them to determine whether they contain nutrients vital to plant growth. Common macronutrients analyzed include compounds of nitrogen, phosphorus, potassium, calcium, magnesium, and sulfur. Soil is also assessed for several micronutrients, like zinc and boron. The percentage of organic matter, soil pH, and nutrient holding capacity (cation exchange capacity) are tested in a regional laboratory. Agronomists will interpret these lab reports and make recommendations to balance soil nutrients for optimal plant growth.

Soil Conservation

In addition, agronomists develop methods to preserve the soil and to decrease the effects of erosion by wind and water. For example, a technique called contour plowing may be used to prevent soil erosion and conserve rainfall. Researchers in agronomy also seek ways to use the soil more effectively in solving other problems. Such problems include the disposal of human and animal manure, water pollution, and pesticide build-up in the soil. Techniques include no-tilling crops, planting of soil-binding grasses along contours on steep slopes, and contour drains of depths up to 1 metre.

Agroecology

Agroecology is the management of agricultural systems with an emphasis on ecological and environmental perspectives. This area is closely associated with work in the areas of sustainable agriculture, organic farming, and alternative food systems and the development of alternative cropping systems.

Theoretical Modeling

Theoretical production ecology tries to quantitatively study the growth of crops. The plant is treated as a kind of biological factory, which processes light, carbon dioxide, water, and nutrients into harvestable parts. Main parameters kept into consideration are temperature, sunlight, standing crop biomass, plant production distribution, nutrient and water supply.

Permaculture

Permaculture is a system of agricultural and social design principles centered on simulating or directly utilizing the patterns and features observed in natural ecosystems. Permaculture was developed, and the term coined by Bill Mollison and David Holmgren in 1968.

It has many branches that include but are not limited to ecological design, ecological engineering, environmental design, construction and integrated water resources management that develops sustainable architecture, regenerative and self-maintained habitat and agricultural systems modeled from natural ecosystems.

Mollison has said: "Permaculture is a philosophy of working with, rather than against nature; of protracted and thoughtful observation rather than protracted and thoughtless labor; and of looking at plants and animals in all their functions, rather than treating any area as a single product system."

History

In 1929, Joseph Russell Smith took up an antecedent term as the subtitle for *Tree Crops: A Permanent Agriculture*, a book in which he summed up his long experience experimenting with fruits and nuts as crops for human food and animal feed. Smith saw the world as an inter-related whole and suggested mixed systems of trees and crops underneath. This book inspired many individuals intent on making agriculture more sustainable, such as Toyohiko Kagawa who pioneered forest farming in Japan in the 1930s.

The definition of permanent agriculture as that which can be sustained indefinitely was supported by Australian P. A. Yeomans in his 1964 book *Water for Every Farm*. Yeomans introduced an observation-based approach to land use in Australia in the 1940s, and the keyline design as a way of managing the supply and distribution of water in the 1950s.

Stewart Brand's works were an early influence noted by Holmgren. Other early influences include Ruth Stout and Esther Deans, who pioneered no-dig gardening, and Masanobu Fukuoka who, in the late 1930s in Japan, began advocating no-till orchards, gardens and natural farming.

Core Tenets and Principles of Design

The three core tenets of permaculture are:

- *Care for the earth*: Provision for all life systems to continue and multiply. This is the first principle, because without a healthy earth, humans cannot flourish.

- *Care for the people*: Provision for people to access those resources necessary for their existence.

- *Return of surplus*: Reinvesting surpluses back into the system to provide for the first two ethics. This includes returning waste back into the system to recycle into usefulness. The third ethic is sometimes referred to as Fair Share to reflect that each of us should take no more than what we need before we reinvest the surplus.

Permaculture design emphasizes patterns of landscape, function, and species assemblies. It determines where these elements should be placed so they can provide maximum benefit to the local environment. The central concept of permaculture is maximizing useful connections between components and synergy of the final design. The focus of permaculture, therefore, is not on each separate element, but rather on the relationships created among elements by the way they are placed together; the whole becoming greater than the sum of its parts. Permaculture design therefore seeks to minimize waste, human labor, and energy input by building systems with maximal benefits between design elements to achieve a high level of synergy. Permaculture designs evolve over time by taking into account these relationships and elements and can become extremely complex systems that produce a high density of food and materials with minimal input.

The design principles which are the conceptual foundation of permaculture were derived from the science of systems ecology and study of pre-industrial examples of sustainable land use. Permaculture draws from several disciplines including organic farming, agroforestry, integrated farming, sustainable development, and applied ecology. Permaculture has been applied most commonly to the design of housing and landscaping, integrating techniques such as agroforestry, natural building, and rainwater harvesting within the context of permaculture design principles and theory.

Theory

Twelve Design Principles

Twelve Permaculture design principles articulated by David Holmgren in his *Permaculture: Principles and Pathways Beyond Sustainability*:

1. *Observe and interact*: By taking time to engage with nature we can design solutions that suit our particular situation.

2. *Catch and store energy*: By developing systems that collect resources at peak abundance, we can use them in times of need.

3. *Obtain a yield*: Ensure that you are getting truly useful rewards as part of the work that you are doing.

4. *Apply self-regulation and accept feedback*: We need to discourage inappropriate activity to ensure that systems can continue to function well.

5. *Use and value renewable resources and services*: Make the best use of nature's abundance to reduce our consumptive behavior and dependence on non-renewable resources.

6. *Produce no waste*: By valuing and making use of all the resources that are available to us, nothing goes to waste.

7. *Design from patterns to details*: By stepping back, we can observe patterns in nature and society. These can form the backbone of our designs, with the details filled in as we go.

8. *Integrate rather than segregate*: By putting the right things in the right place, relationships develop between those things and they work together to support each other.

9. *Use small and slow solutions*: Small and slow systems are easier to maintain than big ones, making better use of local resources and producing more sustainable outcomes.

10. *Use and value diversity*: Diversity reduces vulnerability to a variety of threats and takes advantage of the unique nature of the environment in which it resides.

11. *Use edges and value the marginal*: The interface between things is where the most interesting events take place. These are often the most valuable, diverse and productive elements in the system.

12. *Creatively use and respond to change*: We can have a positive impact on inevitable change by carefully observing, and then intervening at the right time.

Layers

Suburban permaculture garden in Sheffield, UK with different layers of vegetation

Layers are one of the tools used to design functional ecosystems that are both sustainable and of direct benefit to humans. A mature ecosystem has a huge number of relationships between its component parts: trees, understory, ground cover, soil, fungi, insects, and animals. Because plants grow to different heights, a diverse community of life is able to grow in a relatively small space, as the vegetation occupies different layers. There are generally seven recognized layers in a food forest, although some practitioners also include fungi as an eighth layer.

1. The canopy: the tallest trees in the system. Large trees dominate but typically do not saturate the area, i.e. there exist patches barren of trees.

2. Understory layer: trees that revel in the dappled light under the canopy.

3. Shrub layer: a diverse layer of woody perennials of limited height. includes most berry bushes.

4. Herbaceous layer: Plants in this layer die back to the ground every winter (if winters are cold enough, that is). They do not produce woody stems as the Shrub layer does. Many culinary and medicinal herbs are in this layer. A large variety of beneficial plants fall into this layer. May be annuals, biennials or perennials.

5. Soil surface/Groundcover: There is some overlap with the Herbaceous layer and the Groundcover layer; however plants in this layer grow much closer to the ground, grow densely to fill bare patches of soil, and often can tolerate some foot traffic. Cover crops retain soil and lessen erosion, along with green manures that add nutrients and organic matter to the soil, especially nitrogen.

6. Rhizosphere: Root layers within the soil. The major components of this layer are the soil and the organisms that live within it such as plant roots (including root crops such as potatoes and other edible tubers), fungi, insects, nematodes, worms, etc.

7. Vertical layer: climbers or vines, such as runner beans and lima beans (vine varieties).

Guilds

There are many forms of guilds, including guilds of plants with similar functions (that could interchange within an ecosystem), but the most common perception is that of a mutual support guild. Such a guild is a group of species where each provides a unique set of diverse functions that work in conjunction, or harmony. Mutual support guilds are groups of plants, animals, insects, etc. that work well together. Some plants may be grown for food production, some have tap roots that draw nutrients up from deep in the soil, some are nitrogen-fixing legumes, some attract beneficial insects, and others repel harmful insects. When grouped together in a mutually beneficial arrangement, these plants form a guild. See Dave Jacke's work on edible forest gardens for more information on other guilds, specifically resource-partitioning and community-function guilds.

Edge Effect

The edge effect in ecology is the effect of the juxtaposition or placing side by side of contrasting environments on an ecosystem. Permaculturists argue that, where vastly differing systems meet, there is an intense area of productivity and useful connections. An example of this is the coast; where the land and the sea meet there is a particularly rich area that meets a disproportionate percentage of human and animal needs. So this idea is played out in permacultural designs by using spirals in the herb garden or creating ponds that have wavy undulating shorelines rather than a simple circle or oval (thereby increasing the amount of edge for a given area).

Zones

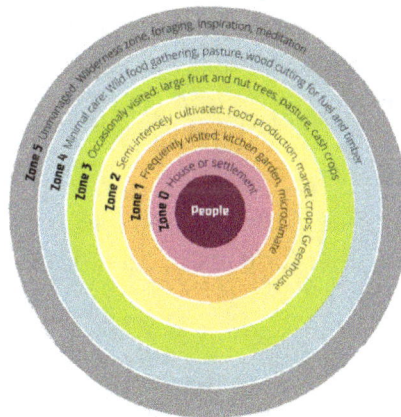

Permaculture Zones 0-5.

Zones are a way of intelligently organizing design elements in a human environment on the basis of the frequency of human use and plant or animal needs. Frequently manipulated or harvested elements of the design are located close to the house in zones 1 and 2. Less frequently used or manipulated elements, and elements that benefit from isolation (such as wild species) are farther away. Zones are about positioning things appropriately, and are numbered from 0 to 5.

Zone 0

> The house, or home center. Here permaculture principles would be applied in terms of aiming to reduce energy and water needs, harnessing natural resources such as sunlight, and generally creating a harmonious, sustainable environment in which to live and work. Zone 0 is an informal designation, which is not specifically defined in Bill Mollison's book.

Zone 1

> The zone nearest to the house, the location for those elements in the system that require frequent attention, or that need to be visited often, such as salad crops, herb plants, soft fruit like strawberries or raspberries, greenhouse and cold frames, propagation area, worm compost bin for kitchen waste, etc. Raised beds are often used in zone 1 in urban areas.

Zone 2

> This area is used for siting perennial plants that require less frequent maintenance, such as occasional weed control or pruning, including currant bushes and orchards, pumpkins, sweet potato, etc. This would also be a good place for beehives, larger scale composting bins, and so on.

Zone 3

> The area where main-crops are grown, both for domestic use and for trade purposes. After establishment, care and maintenance required are fairly minimal (provided mulches and similar things are used), such as watering or weed control maybe once a week.

Zone 4

> A semi-wild area. This zone is mainly used for forage and collecting wild food as well as production of timber for construction or firewood.

Zone 5

> A wilderness area. There is no human intervention in zone 5 apart from the observation of natural ecosystems and cycles. Through this zone we build up a natural reserve of bacteria, moulds and insects that can aid the zones above it.

People and Permaculture

Permaculture uses observation of nature to create regenerative systems, and the place where this has been most visible has been on the landscape. There has been a growing awareness though that firstly, there is the need to pay more attention to the peoplecare ethic, as it is often the dynamics of people that can interfere with projects, and secondly that the principles of permaculture can be used as effectively to create vibrant, healthy and productive people and communities as they have been in landscapes.

Domesticated Animals

Domesticated animals are often incorporated into site design.

Common Practices

Agroforestry

Agroforestry is an integrated approach of using the interactive benefits from combining trees and shrubs with crops and/or livestock. It combines agricultural and forestry technologies to create more diverse, productive, profitable, healthy and sustainable land-use systems. In agroforestry systems, trees or shrubs are intentionally used within agricultural systems, or non-timber forest products are cultured in forest settings.

Forest gardening is a term permaculturalists use to describe systems designed to mimic natural forests. Forest gardens, like other permaculture designs, incorporate processes and relationships that the designers understand to be valuable in natural ecosystems. The terms forest garden and food forest are used interchangeably in the permaculture literature. Numerous permaculturists are proponents of forest gardens, such as Graham Bell, Patrick Whitefield, Dave Jacke, Eric Toensmeier and Geoff Lawton. Bell started building his forest garden in 1991 and wrote the book *The Permaculture Garden* in 1995, Whitefield wrote the book *How to Make a Forest Garden* in 2002, Jacke and Toensmeier co-authored the two volume book set *Edible Forest Gardening* in 2005, and Lawton presented the film *Establishing a Food Forest* in 2008.

Tree Gardens, such as Kandyan tree gardens, in South and Southeast Asia, are often hundreds of years old. Whether they derived initially from experiences of cultivation and forestry, as is the case in agroforestry, or whether they derived from an understanding of forest ecosystems, as is the case for permaculture systems, is not self-evident. Many studies of these systems, especially those that predate the term permaculture, consider these systems to be forms of agroforestry. Permaculturalists who include existing and ancient systems of polycropping with woody species as examples of food forests may obscure the distinction between permaculture and agroforestry.

Food forests and agroforestry are parallel approaches that sometimes lead to similar designs.

Hügelkultur

Hügelkultur is the practice of burying large volumes of wood to increase soil water retention. The porous structure of wood acts as a sponge when decomposing underground. During the rainy season, masses of buried wood can absorb enough water to sustain crops through the dry season. This technique has been used by permaculturalists Sepp Holzer, Toby Hemenway, Paul Wheaton, and Masanobu Fukuoka.

Natural Building

A natural building involves a range of building systems and materials that place major emphasis on sustainability. Ways of achieving sustainability through natural building focus on durability and the use of minimally processed, plentiful or renewable resources, as well as those that, while recycled or salvaged, produce healthy living environments and maintain indoor air quality.

The basis of natural building is the need to lessen the environmental impact of buildings and other supporting systems, without sacrificing comfort, health or aesthetics. To be more sustainable, natural building uses primarily abundantly available, renewable, reused or recycled materials. In addition to relying on natural building materials, the emphasis on the architectural design is heightened. The orientation of a building, the utilization of local climate and site conditions, the emphasis on natural ventilation through design, fundamentally lessen operational costs and positively impact the environment. Building compactly and minimizing the ecological footprint is common, as are on-site handling of energy acquisition, on-site water capture, alternate sewage treatment and water reuse.

Rainwater Harvesting

Rainwater harvesting is the accumulating and storing of rainwater for reuse before it reaches the aquifer. It has been used to provide drinking water, water for livestock, water for irrigation, as well as other typical uses. Rainwater collected from the roofs of houses and local institutions can make an important contribution to the availability of drinking water. It can supplement the subsoil water level and increase urban greenery. Water collected from the ground, sometimes from areas which are especially prepared for this purpose, is called stormwater harvesting.

Greywater is wastewater generated from domestic activities such as laundry, dishwashing, and bathing, which can be recycled on-site for uses such as landscape irrigation and constructed wetlands. Greywater is largely sterile, but not potable (drinkable). Greywater differs from water from the toilets which is designated sewage or blackwater, to indicate it contains human waste. Blackwater is septic or otherwise toxic and cannot easily be reused. There are, however, continuing efforts to make use of blackwater or human waste. The most notable is for composting through a process known as humanure; a combination of the words human and manure. Additionally, the methane in humanure can be collected and used similar to natural gas as a fuel, such as for heating or cooking, and is commonly referred to as biogas. Biogas can be harvested from the human waste and the remainder still used as humanure. Some of the simplest forms of humanure use include a composting toilet or an outhouse or dry bog surrounded by trees that are heavy feeders which can be coppiced for wood fuel. This process eliminates the use of a standard toilet with plumbing.

Sheet Mulching

In agriculture and gardening, mulch is a protective cover placed over the soil. Any material or combination can be used as mulch, such as stones, leaves, cardboard, wood chips, gravel, etc., though in permaculture mulches of organic material are the most common because they perform more functions. These include: absorbing rainfall, reducing evaporation, providing nutrients, increasing organic matter in the soil, feeding and creating habitat for soil organisms, suppressing weed growth and seed germination, moderating diurnal temperature swings, protecting against frost, and reducing erosion. Sheet mulching is an agricultural no-dig gardening technique that attempts to mimic natural processes occurring within forests. Sheet mulching mimics the leaf cover that is found on forest floors. When deployed properly and in combination with other Permacultural principles, it can generate healthy, productive and low maintenance ecosystems.

Sheet mulch serves as a "nutrient bank," storing the nutrients contained in organic matter and slowly making these nutrients available to plants as the organic matter slowly and naturally breaks down. It also improves the soil by attracting and feeding earthworms, slaters and many other soil micro-organisms, as well as adding humus. Earthworms "till" the soil, and their worm castings are among the best fertilizers and soil conditioners. Sheet mulching can be used to reduce or eliminate undesirable plants by starving them of light, and can be more advantageous than using herbicide or other methods of control.

Intensive Rotational Grazing

Grazing has long been blamed for much of the destruction we see in the environment. However, it has been shown that when grazing is modeled after nature, the opposite effect can be seen. Also known as cell grazing, managed intensive rotational grazing (MIRG) is a system of grazing in which ruminant and non-ruminant herds and/or flocks are regularly and systematically moved to fresh pasture, range, or forest with the intent to maximize the quality and quantity of forage growth. This disturbance is then followed by a period of rest which allows new growth. MIRG can be used with cattle, sheep, goats, pigs, chickens, rabbits, geese, turkeys, ducks and other animals depending on the natural ecological community that is being mimicked. Sepp Holzer and Joel Salatin have shown how the disturbance caused by the animals can be the spark needed to start ecological succession or prepare ground for planting. Allan Savory's holistic management technique has been likened to "a permaculture approach to rangeland management". One variation on MIRG that is gaining rapid popularity is called eco-grazing. Often used to either control invasives or re-establish native species, in eco-grazing the primary purpose of the animals is to benefit the environment and the animals can be, but are not necessarily, used for meat, milk or fiber.

Keyline Design

Keyline design is a technique for maximizing beneficial use of water resources of a piece of land developed in Australia by farmer and engineer P. A. Yeomans. The *Keyline* refers to a specific topographic feature linked to water flow which is used in designing the drainage system of the site.

Fruit Tree Management

> The no-pruning option is usually ignored by fruit experts, though often practised by default in people's back gardens! But it has its advantages. Obviously it reduces work, and more surprisingly it can lead to higher overall yields.
>
> — *Whitefield, Patrick, How to make a forest garden, p. 16*

Masanobu Fukuoka, as part of early experiments on his family farm in Japan, experimented with no-pruning methods, noting that he ended up killing many fruit trees by simply letting them go, which made them become convoluted and tangled, and thus unhealthy. Then he realised this is the difference between natural-form fruit trees and the process of change of tree form that results from abandoning previously-pruned unnatural fruit trees. He concluded that the trees should be raised all their lives without pruning, so they form healthy and efficient branch patterns that follow their natural inclination. This is part of his implementation of the Tao-philosophy of Wú wéi translated in part as no-action (against nature), and he described it as no unnecessary pruning, nature farming or "do-nothing" farming, of fruit trees, distinct from non-intervention or literal no-pruning. He ultimately achieved yields comparable to or exceeding standard/intensive practices of using pruning and chemical fertilisation.

Another proponent of the no, or limited, pruning method is Sepp Holzer who used the method in connection with Hügelkultur berms. He has successfully grown several varieties of fruiting trees at altitudes (approximately 9,000 feet (2,700 m)) far above their normal altitude, temperature, and snow load ranges. He notes that the Hügelkultur berms kept and/or generated enough heat to allow the roots to survive during alpine winter conditions. The point of having unpruned branches, he notes, was that the longer (more naturally formed) branches bend over under the snow load until they touched the ground, thus forming a natural arch against snow loads that would break a shorter, pruned, branch.

Compost Management

Compost Basket

Compost Basket is a way to the permanent management of compost materials. The idea of Compost Basket comes from Gyulai Iván. He invented and used it in the Gömörszőlős Educational Center. The inner circle is 1 meter deep. This is the place where compost material are put in. Under this is the outer circle, 40 cm deep. Here the nutrients come down.

Mollison and Holmgren

Bill Mollison in January 2008.

In the late 1960s, Bill Mollison and David Holmgren started developing ideas about stable agricultural systems on the southern Australian island state of Tasmania. This was a result of the danger of the rapidly growing use of industrial-agricultural methods. In their view, these methods were highly dependent on non-renewable resources, and were additionally poisoning land and water, reducing biodiversity, and removing billions of tons of topsoil from previously fertile landscapes. A design approach called *permaculture* was their response and was first made public with the publication of their book *Permaculture One* in 1978.

By the early 1980s, the concept had broadened from agricultural systems design towards sustainable human habitats. After *Permaculture One*, Mollison further refined and developed the ideas by designing hundreds of permaculture sites and writing more detailed books, notably *Permaculture: A Designers Manual*. Mollison lectured in over 80 countries and taught his two-week Permaculture Design Course (PDC) to many hundreds of students.Mollison "encouraged graduates to become teachers themselves and set up their own institutes and demonstration sites. This multiplier effect was critical to permaculture's rapid expansion."

In 1991, a four-part television documentary by ABC productions called "The Global Gardener" showed permaculture applied to a range of worldwide situations, bringing the concept to a much broader public. In 2012, the UMass Permaculture Initiative won the White House "Champions of Change" sustainability contest, which declared that "they demonstrate how permaculture can feed a growing population in an environmentally sustainable and socially responsible manner".

In 1997, Holmgren explained that the primary agenda of the permaculture movement is to assist people to become more self-reliant through the design and development of productive and sustainable gardens and farms.

In 2014, Holmgren endorsed and helped launch a new Australian permaculture magazine, Pip Magazine.

Notable Permaculturists

Joseph Russell Smith took up an antecedent term as the subtitle for *Tree Crops: A Permanent Agriculture*, a book in which he summed up his long experience experimenting with fruits and nuts as crops for human food and animal feed. By that year (1929), Smith saw the world as an interrelated whole and suggested mixed systems of trees and crops underneath. This book, along with the work of P. A. Yeomans (the latter created the keyline design concept), inspired many individuals intent on making permaculture a valid means of sustainable food production. Bill Mollison and David Holmgren developed it further, and permaculturists were trained under the umbrella of Bill Mollison's train the trainer system.

Among some of the more recognizable names who received their original training within Mollison's PDC system would include Geoff Lawton and Toby Hemenway, each of whom have more than 25 years experience teaching and promoting permaculture as a sustainable way of growing food. Simon Fjell was a Founding Director of the Permaculture Institute in late 1979, over 40 years experience, having first met Mollison in 1976. He has since worked in every continent.

The permaculture movement also spread throughout Asia and Central America, with Hong Kong-based Asian Institute of Sustainable Architecture (AISA), Rony Lec leading the foundation of the Mesoamerican Permaculture Institute (IMAP) in Guatemala and Juan Rojas co-founding the Permaculture Institute of El Salvador.

Trademark and Copyright Issues

There has been contention over who, if anyone, controls legal rights to the word *perma-culture*: is it trademarked or copyrighted? and if so, who holds the legal rights to the use of the word? For a long time Bill Mollison claimed to have copyrighted the word, and his books said on the copyright page, "The contents of this book and the word PERMA-CULTURE are copyright." These statements were largely accepted at face-value within the permaculture community. However, copyright law does not protect names, ideas, concepts, systems, or methods of doing something; it only protects the expression or the description of an idea, not the idea itself. Eventually Mollison acknowledged that he was mistaken and that no copyright protection existed for the word *permaculture*.

In 2000, Mollison's US based Permaculture Institute sought a service mark (a form of trademark) for the word *permaculture* when used in educational services such as con-ducting classes, seminars, or workshops. The service mark would have allowed Molli-son and his two Permaculture Institutes (one in the US and one in Australia) to set en-forceable guidelines regarding how permaculture could be taught and who could teach it, particularly with relation to the PDC, despite the fact that he had instituted a system of certification of teachers to teach the PDC in 1993. The service mark failed and was abandoned in 2001. Also in 2001 Mollison applied for trademarks in Australia for the terms "Permaculture Design Course" and "Permaculture Design". These applications were both withdrawn in 2003. In 2009 he sought a trademark for "Permaculture: A Designers' Manual" and "Introduction to Permaculture", the names of two of his books. These applications were withdrawn in 2011. There has never been a trademark for the word *permaculture* in Australia.

Criticisms

General Criticisms

In 2011, Owen Hablutzel argued that "permaculture has yet to gain a large amount of specific mainstream scientific acceptance," and that "the sensitiveness to being per-ceived and accepted on scientific terms is motivated in part by a desire for permacul-ture to expand and become increasingly relevant." Bec-Hellouin permaculture farm engaged in a research program in partnership with INRA and AgroParisTech to collect scientific data.

In his books *Sustainable Freshwater Aquaculture* and *Farming in Ponds and Dams*, Nick Romanowski expresses the view that the presentation of aquaculture in Bill Mol-lison's books is unrealistic and misleading.

Agroforestry

Greg Williams argues that forests cannot be more productive than farmland because the net productivity of forests decline as they mature due to ecological succession. Proponents of permaculture respond that this is true only if one compares data between woodland forest and climax vegetation, but not when comparing farmland vegetation with woodland forest. For example, ecological succession generally results in a forest's productivity rising after its establishment only until it reaches the *woodland state* (67% tree cover), before declining until *full maturity*.

Polyculture

Polyculture providing useful within-field diversity: companion planting of carrots and onions. The onion smell puts off carrot root fly, while the smell of carrots puts off onion fly.

Polyculture is agriculture using multiple crops in the same space, providing crop diversity in imitation of the diversity of natural ecosystems, and avoiding large stands of single crops, or monoculture. It includes multi-cropping, intercropping, companion planting, beneficial weeds, and alley cropping. It is the raising at the same time and place of more than one species of plant or animal. Polyculture is one of the principles of permaculture.

Advantages

Polyculture, though it often requires more labor, has two main advantages over monoculture.

Polyculture reduces susceptibility to disease. For example, a study in China showed that planting several varieties of rice in the same field increased yields by 89%, largely because of a dramatic (94%) decrease in the incidence of disease, which made pesticides redundant.

Polyculture increases local biodiversity. This is one example of reconciliation ecology, or accommodating biodiversity within human landscapes. This may also form part of a biological pest control program.

Silviculture

Silviculture is the practice of controlling the establishment, growth, composition, health, and quality of forests to meet diverse needs and values.

The name comes from the Latin *silvi-* (forest) + culture (as in growing). The study of forests and woods is termed silvology. Silviculture also focuses on making sure that the treatment(s) of forest stands are used to preserve and to better their productivity.

Generally, silviculture is the science and art of growing and tending forest crops, based on a knowledge of silvics, i.e., the study of the life history and general characteristics of forest trees and stands, with particular reference to locality factors. More particularly, silviculture is the theory and practice of controlling the establishment, composition, constitution, and growth of forests. No matter how forestry as a science is constituted, the kernel of the business of forestry is silviculture, as it includes direct action in the forest, and in it all economic objectives and technical considerations ultimately converge. The kernel of silviculture is regeneration.

Suggestions for how best to go about the job, presented by Jeglum et al. (2003), though aimed primarily at the boreal forest in Ontario, merit wider consideration. The 110-page publication describes Best Management Practices, first by general principles, then by sensitive sites. Illustrations are plentiful and are well chosen to complement this excellent text.

To some the distinction between forestry and silviculture is that silviculture is applied at the stand level and forestry is broader. For example, John D. Matthews says "complete regimes for regenerating, tending, and harvesting forests" are called "silvicultural systems".

So adaptive management is required for silviculture, whereas forestry can be natural, conserved land without a stand level treatment being applied. A common taxonomy divides silviculture into regenerating, tending and harvesting techniques.

An employee of the White Mountain Apache Tribe forestry department uses a hoedad to dig a hole for a ponderosa pine seedling in the shade of a burned tree stump, which protects the seedling from windy conditions that dry the trees

Silvicultural Systems

The origin of forestry in German speaking Europe has defined silvicultural systems broadly as Hochwald - High forest, Mittelwald - Coppice with standards (and Compound coppice), Short rotation coppice and Niederwald - Coppice. There are other systems too. These varied silvicultural systems include several harvesting methods, which are often wrongly said to be a silvicultural systems, but may also be called as rejuvenating or regenerating method depending on the purpose.

Hochwald is further subdivided in German into: Altersklassenwald - Age class forest which includes:

- *even aged forestry*

    o   - Kahlschlag - Clear cutting

    o   - Schirmschlag - Shelterwood cutting

    o   - Seed-tree method

- *un-even aged forestry*

    o   - Femelschlag (Femelschlagverfahren) - The Femel selection cutting (Group selection cutting),

    o   - Saumschlag (Saumfemelschlagbetrieb) - Strip selection cutting (strip-and-group felling system)

    o   - Schirmkeilschlag - Shelterwood wedge cutting

    o   - Mischformen - mixed form regeneration methods.

and Dauerwald - Continuous cover forestry, in which are found:

- *un-even aged forestry*

    o   - Plenterwald - Selection forest

    o   - Zielstärkennutzung - Target diameter harvesting.

From these names the impression is given that these are neatly defined systems, but in practice there are a variety of variations within these harvesting methods according to local ecology and site conditions. While an individual location can be identified where it could be said to be the archetypal form of harvesting technique (they all originated somewhere with a particular forester, and have been described in the scientific literature), and broad generalizations can be made these are merely rules of thumb rather than strict blue prints on how techniques might be applied. This misunderstanding has meant that many older English textbooks did not capture the true complexity of silviculture as practiced where it originated in Mitteleuropa.

This silviculture was culturally predicated on wood production in temperate and boreal climates and did not deal with tropical forestry. The misapplication of this philosophy to those tropical forests has been problematic. There is also an alternative silvicultural tradition which developed in Japan and thus created a different biocultural landscape called satoyama.

After harvesting comes regeneration, which may be split into Natural and Artificial. And tending which includes release treatments, pruning, thinning and intermediate treatments. As it can be imagined it maybe that any of these 3 phases (harvesting - regeneration - tending) may happen at the same time, within a stand, depending on the goal for that particular stand.

Regeneration

Regeneration is basic to the continuation of forested, as well as to the afforestation of tree-less land. Regeneration can take place through self-sown seed ("natural regeneration"), by artificially sown seed, or by planted seedlings. In whichever case, the performance of regeneration depends on its growth potential and the degree to which its environment allows the potential to be expressed. Seed, of course, is needed for all regeneration modes, both for natural or artificial sowing and for raising planting stock in a nursery.

Natural regeneration is:

"Human-assisted natural regeneration" means establishment of a forest age class from natural seeding or sprouting in an area after harvesting in that area through selection cutting, shelter (or seed-tree) harvest, soil preparation, or restricting the size of a clear-cut stand to secure natural regeneration from the surrounding trees.

The process of natural regeneration involves the renewal of forests by means of self-sown seeds, root suckers, or coppicing. In natural forests, conifers rely almost entirely on regeneration through seed. Most of the broadleaves, however, are able to regenerate by the means of emergence of shoots from stumps (coppice) and broken stems.

Seedbed Requirements

Any seed, self-sown or artificially applied, requires a seedbed suitable for securing germination.

In order to germinate, a seed requires suitable conditions of temperature, moisture, and aeration. For seeds of many species, light is also necessary, and facilitates the germination of seeds in other species, but spruces are not exacting in their light requirements, and will germinate without light. White spruce seed germinated at 35 °F (1.7 °C) and 40 °F (4.4 °C) after continuous stratification for one year or longer and developed radicles < 6 cm long in the cold room. When exposed to light, those germinants developed chlorophyll and were normally phototropic with continued elongation.

For survival in the short and medium terms, a germinant needs: a continuing supply of moisture; freedom from lethal temperature; enough light to generate sufficient photosynthate to support respiration and growth, but not enough to generate lethal stress in the seedling; freedom from browsers, tramplers, and pathogens; and a stable root system. Shade is very important to the survival of young seedlings. In the longer term, there must be an adequate supply of essential nutrients and an absence of smothering.

In undisturbed forest, decayed windfallen stemwood provides the most favourable seedbed for germination and survival, moisture supply being dependable, and the elevation of seedlings somewhat above the general level of the forest floor reduces the danger of smothering by leaves and snow-pressed minor vegetation; nor is such a microsite likely to be subject to flooding. Advantages conferred by those microsites include: more light, higher temperatures in the rooting zone, and better mycorrhizal development. In a survey in the Porcupine Hills, Manitoba, 90% of all spruce seedlings were rooted in rotten wood.

Mineral soil seedbeds are more receptive than the undisturbed forest floor, and are generally moister and more readily rewetted than the organic forest floor. However, exposed mineral soil, much more so than organic-surfaced soil, is subject to frost heaving and shrinkage during drought. The forces generated in soil by frost or drought are quite enough to break roots.

The range of microsites occurring on the forest floor can be broadened, and their frequency and distribution influenced by site preparation. Each microsite has its own microclimate. Microclimates near the ground are better characterized by vapour pressure deficit and net incident radiation, rather than the standard measurements of air temperature, precipitation, and wind pattern.

Aspect is an important component of microclimate, especially in relation to temperature and moisture regimes. Germination and seedling establishment of Engelmann spruce were much better on north than on south aspect seedbeds in the Fraser Experimental Forest, Colorado; the ratios of seeds to 5-year-old seedlings were determined as 32:1, 76:1, and 72:1 on north aspect bladed-shaded, bladed-unshaded, and undisturbed-shaded seedbeds, respectively. Clearcut openings of 1.2 ha to 2.0 ha adjacent to an adequate seed source, and not more than 6 tree heights wide, could be expected to secure acceptable regeneration (4,900, 5-year-old trees/ha), whereas on undisturbed-unshaded north aspects, and on all seedbed treatments tested on south aspects, seed to seedling ratios were so high that the restocking of any clearcut opening would be questionable.

At least 7 variable factors may influence seed germination: (1) seed characteristics, (2) light, (3) oxygen, (4) soil reaction (ph), (5) temperature, (6) moisture, and (7) seed enemies. Moisture and temperature are the most influential, and both are affected by exposure. The difficulty of securing natural regeneration of Norway spruce and Scots pine in northern Europe led to the adoption of various forms of reproduction cuttings that provided partial shade or protection to seedlings from hot sun and wind. The main objective of echeloned strips or border-cuttings with northeast exposure was to protect regeneration from overheating, and was originated in Germany and deployed success-fully by Alarik (1925) and others in Sweden. On south and west exposures, direct inso-lation and heat reflected from tree trunks often result in temperatures lethal to young seedlings, as well as desiccation of the surface soil, which inhibits germination. The sun is less injurious on eastern exposures because of the lower temperature in the early morning, related to higher humidity and presence of dew.

Baldwin (1933), after noting that summer temperatures in North America are often higher than those in places where border-cuttings have been found useful, reported the results of a survey of regeneration in a stand of red spruce plus scattered white spruce that had been isolated by clearcutting on all sides, so furnishing an opportunity for observing regeneration on different exposures in this old-field stand at Dummer, New Hampshire. The regeneration included a surprisingly large number of balsam fir seed-lings from the 5% stand component of that species. The maximum density of spruce regeneration, determined 4 rods (20 m) inside from the edge of the stand on a North 20°E exposure, was 600,000/ha, with almost 100,000 balsam fir seedlings.

A prepared seedbed remains receptive for a relatively short period, seldom as long as 5 years, sometimes as short as 3 years. Seedbed receptivity on moist, fertile sites decreas-es with particular rapidity, and especially on such sites, seedbed preparation should be scheduled to take advantage of good seed years. In poor seed years, site preparation can be carried out on mesic and drier sites with more chance of success, because of the gen-erally longer receptivity of seedbeds there than those on moister sites. Although an in-different seed year can suffice if seed distribution is good and environmental conditions favourable to seedling germination and survival, small amounts of seed are particularly vulnerable to depredation by small mammals. Considerable flexibility is possible in timing site preparation to coincide with cone crops. Treatment can be apllied either before any logging takes place, between partial cuts, or after logging. In cut and leave strips, seedbed preparation can be carried out as a single operation, pre-scarifying the leave strips, post-scarifying the cut strips.

Broadcast burning is not recommended as a method of preparing sites for natural re-generation, as it rarely exposes enough mineral soil to be sufficiently receptive, and the charred organic surfaces are a poor seedbed for spruce. A charred surface may get too hot for good germination and may delay germination until fall, with subsequent over-winter mortality of unhardened seedlings. Piling and burning of logging slash, howev-er, can leave suitable exposures of mineral soil.

Season of Planting

Artificial Regeneration

With a view to reducing the time needed to produce planting stock, experiments were carried out with white spruce and three other coniferous species from Wisconsin seed in the longer, frost-free growing season in Florida, 125 vs. 265 days in central Wisconsin and northern Florida, respectively. As the species studied are adapted to long photoperiods, extended daylengths of 20 hours were applied in Florida. Other seedlings were grown under extended daylength in Wisconsin and with natural daylength in both areas. After two growing seasons, white spruce under long days in Florida were about the same as those in Wisconsin, but twice as tall as plants under natural Wisconsin photoperiods. Under natural days in Florida, with the short local photoperiod, white spruce was severely dwarfed and had a low rate of survival. Black spruce responded similarly. After two growing seasons, long day plants of all 4 species in Florida were well balanced, with good development of both roots and shoots, equaling or exceeding the minimum standards for 2+1 and 2+2 outplanting stock of Lake States species. Their survival when lifted in February and outplanted in Wisconsin equalled that of 2+2 Wisconsin-grown transplants. Artificial extension of the photoperiod in the northern Lake States greatly increased height increment of white and black spruces in the second growing season.

Optimum conditions for seedling growth have been determined for the production of containerized planting stock. Alternating day/night temperatures have been found more suitable than a constant temperature; at 400 lumens/m² light regime, a 28 °C/20 °C day/night temperatures have been recommended for white spruce. However, temperature optima are not necessarily the same at different ages and sizes. Tinus (1984) investigated the effects of combinations of day and night temperature on height, caliper, and dry weight of 4 seed sources of Engelmann spruce. The 4 seed sources appeared to have very similar temperature requirements, with night optima about the same of slightly lower than daylight optima.

Tree provenance is important in artificial regeneration. Good provenance takes into account suitable tree genetics and a good environmental fit for planted / seeded trees in a forest stand. The wrong genotype can lead to failed regeneration, or poor trees that are prone to pathogens and undesired outcomes.

Artificial regeneration has been a more common method involving planting because it is more dependable than natural regeneration. Planting can involve using seedlings (from a nursery), (un)rooted cuttings, or seeds.

Whichever method is chosen it can be assisted by tending techniques also known as intermediate stand treatments.

The fundamental genetic consideration in artificial regeneration is that seed and planting stock must be adapted to the planting environment. Most commonly, the method of managing seed and stock deployment is through a system of defined seed zones, within which seed and stock can be moved without risk of climatic maladaptation. Ontario adopted a seed zone system in the 1970s based on Hills' (1952) site regions and provincial resource district boundaries, but Ontario's seed zones are now based on homogeneous climatic regions developed with the Ontario Climate Model. The regulations stipulate that source-identified seedlots may be either a general collection, when only the seed zone of origin is known, or a stand collection from a specific latitude and longitude. The movement of general-collection seed and stock across seed zone boundaries is prohibited, but the use of stand-collection seed and stock in another seed zone is acceptable when the Ontario Climate Model shows that the planting site and place of seed origin are climatically similar. The 12 seed zones for white spruce in Quebec are based mainly on ecological regions, with a few modifications for administrative convenience.

Seed quality varies with source. Seed orchards produce seed of the highest quality, then, in order of decreasing seed quality produced, seed production areas and seed collection areas follow, with controlled general collections and uncontrolled general collections producing the least characterized seed.

Seeds

Dewinging, Extraction

When seed is first separated from cones it is mixed with foreign matter, often 2 to 5 times the volume of the seed. The more or less firmly attached membranous wings on the seed must be detached before it is cleaned of foreign matter. The testa must not incur damage during the dewinging process. Two methods have been used, dry and wet. Dry seed may be rubbed gently through a sieve that has a mesh through which only seed without wings can pass. Large quantities of seed can be processed in dewinging machines, which use cylinders of heavy wire mesh and rapidly revolving stiff brushes within to remove the wings. In the wet process, seed with wings attached are spread out 10 cm to 15 cm deep on a tight floor and slightly moistened throughout; light leather flails are used to free seed from the wings. Wang (1973) described a unique wet dewinging procedure using a cement mixer, used at the Petawawa tree seed processing facility. Wings of white and Norway spruce seed can be removed by dampening the seed slightly before it is run through a fanning mill for the last time. Any moistened seed must be dried before fermentation or moulding sets in.

Seed Viability

A fluorescein diacetate (FDA) biochemical viability test for several species of conifer seed, including white spruce, estimates the proportion of live seed (viability) in a seedlot, and hence the percentage germination of a seedlot. The accuracy of predicting percentage germination was within +/- 5 for most seedlots. White spruce seed can be tested for viability by an indirect method, such as the fluorescein diacetate (FDA) test or 'Ultra-sound'; or by the direct growth method of 'germination'. Samples of white spruce seed inspected by Toumey and Stevens (1928) varied in viability from 50% to 100%, but averaged 93%. Rafn (1915) reported 97% viability for white spruce seed.

Germinative Testing

The results of a germination test are commonly expressed as *germinative capacity* or a *germination percentage*, which is the percentage of seeds that germinate during a period of time, ending when germination is practically complete. During extraction and processing, white spruce seeds gradually lost moisture, and total germination increased. Mittal et al. (1987) reported that white spruce seed from Algonquin Park, Ontario, obtained the maximum rate (94% in 6 days) and 99% total germination in 21 days after 14-week pre-chilling. The pre-treatment of 1% sodium hypochlorite increased germinability.

Encouraged by Russian success in using ultrasonic waves to improve the germinative energy and percentage germination of seeds of agricultural crops, Timonin (1966) demonstrated benefits to white spruce germination after exposure of seeds to 1, 2, or 4 minutes of ultrasound generated by an M.S.E. ultrasonic disintegrator with a power consumption of 280 VA and power impact of 1.35 amperes. However, no seeds germinated after 6 minutes of exposure to ultrasound.

Seed Dormancy

Seed dormancy is a complex phenomenon and is not always consistent within species. Cold stratification of white spruce seed to break dormancy has been specified as a requirement, but Heit (1961) and Hellum (1968) regarded stratification as unnecessary. Cone handling and storage conditions affect dormancy in that cold, humid storage (5 °C, 75% to 95% relative humidity) of the cones prior to extraction seemingly eliminated dormancy by overcoming the need to stratify. Periods of cold, damp weather during the period of cone storage might provide natural cold (stratification) treatment. Once dormancy was removed in cone storage, subsequent kiln-drying and seed storage did not reactivate dormancy.

Haddon and Winston (1982) found a reduction in viability of stratified seeds after 2 years of storage and suggested that stress might have been caused by stratification, e.g., by changes in seed biochemistry, reduced embryo vigor, seed aging or actual damage to the embryo. They further questioned the quality of the 2-year-old seed even though high germination occurred in the samples that were not stratified.

Cold Stratification

Cold stratification is the term applied to the storing of seeds in (and, strictly, in layers with) a moist medium, often peat or sand, with a view to maintaining viability and overcoming dormancy. Cold stratification is the term applied to storage at near-freezing temperatures, even if no medium is used. A common method of cold stratification, is to soak seed in tap water for up to 24 h, superficially dry it, then store moist for some weeks or even months at temperatures just above freezing. Although Hellum (1968) found that cold stratification of an Alberta seed source led to irregular germination, with decreasing germination with increasing length of the stratification period, Hocking's (1972) paired test with stratified and nonstratified Alberta seed from several sources revealed no trends in response to stratification. Hocking suggested that seed maturity, handling, and storage needed to be controlled before the need for stratification could be determined. Later, Winston and Haddon (1981) found that the storage of white spruce cones for 4 weeks at 5 °C prior to extraction obviated the need for stratification.

Seed Ripeness

Seed maturity cannot be predicted accurately from cone flotation, cone moisture content, cone specific gravity; but the province of B.C. found embryo occupying 90% + of the corrosion cavity and megagametophyte being firm and whitish in colour are the best predictors for white spruce in B.C., and Quebec can forecast seed maturity some weeks in advance by monitoring seed development in relation to heat-sums and the phenological progression of the inflorescence of fireweed (*Epilobium angustifolium* L.), an associated plant species. Cone collection earlier than one week before seed maturity would reduce seed germination and viability during storage. Four stages of maturation were determined by monitoring carbohydrates, polyols, organic acids, respiration, and metabolic activity. White spruce seeds require a 6-week post-harvest ripening period in the cones to obtain maximum germinability, however, based on cumulative degree-days, seed from the same trees and stand showed 2-week cone storage was sufficient.

Forest Tree Plantations

Plantation Establishment Criteria

Plantations may be considered successful when outplant performance satisfies certain criteria. The term "free growing" is applied in some jurisdictions. Ontario's "Free-to-Grow" (FTG) equivalent relates to a forest stand that meets a minimum stocking standard and height requirement, and is essentially free of competition from surrounding vegetation that might impede growth. The FTG concept was introduced with the advent of the Forest Management Agreement program in Ontario in 1980 and became applicable to all management units in 1986. Policy, procedures, and methodologies readily applicable by forest unit managers to assess the effectiveness of regeneration programs were still under development during the Class Environmental Assessment hearings.

In British Columbia, the Forest Practices Code (1995) governs performance criteria. To minimize the subjectivity of assessing deciduous competition as to whether or not a plantation is established, minimum specifications of number, health, height, and competition have been specified in British Columbia. However, minimum specifications are still subjectively set and may need to be fine-tuned in order to avoid unwarranted delay in according established status to a plantation. For example, a vigorous white spruce with a strong, multi-budded leading shoot and its crown fully exposed to light on 3 sides would not qualify as free-growing in the current British Columbia Code but would hardly warrant description as unestablished.

Competition

Competition arises when individual organisms are sufficiently close together to incur growth constraint through mutual modification of the local environment. Plants may compete for light, moisture and nutrients, but seldom for space *per se*. Vegetation management directs more of the site's resources into usable forest products, rather than just eliminating all competing plants. Ideally, site preparation ameliorates competition to levels that relieve the outplant of constraints severe enough to cause prolonged check.

The diversity of boreal and sub-boreal broadleaf-conifer mixed tree species stands, commonly referred to as the "mixedwoods", largely preclude the utility of generalizations and call for the development of management practices incorporating the greater inherent complexity of broadleaf-conifer mixtures, relative to single-species or mixed-species conifer forest. After harvesting or other disturbance, mixedwood stands commonly enter a prolonged period in which hardwoods overtop the coniferous component, subjecting them to intense competition in an understorey. It is well established that the regeneration and growth potential of understorey conifers in mixedwood stands is correlated to the density of competing hardwoods. To help apply "free-to-grow" regulations in British Columbia and Alberta, management guidelines based on distance-dependent relations within a limited radius of crop trees were developed, but Lieffers et al. (2002) found that free-growing stocking standards did not adequately characterize light competition between broadleaf and conifer components in boreal mixedwood stands, and further noted that adequate sampling using current approaches would be operationally prohibitive.

Many promising plantations have failed through lack of tending. Young crop trees are often ill-equipped to fight it out with competition resurgent following initial site preparation and planting.

Perhaps the most direct evaluation of the effect of competition on plantation establishment is provided by an effective herbicide treatment. The fact that herbicide treatment does not always produce positive results should not obscure the demonstrated potential of herbicides for significantly promoting plantation establishment. Factors that can vitiate the effectiveness of a herbicide treatment include: weather, especially temperature, prior to and during application; weather, especially wind, during application; weather, especially precipitation, in the 12 to 24 hours after application; vegetation characteristics, including species, size, shape, phenological stage, vigour, and distribution of weeds; crop characteristics, including species, phenology, and condition; the effects of other treatments, such as preliminary shearblading, burning or other prescribed or accidental site preparation; and the herbicide used, including dosage, formulation, carrier, spreader, and mode of application. There is a lot that can go wrong, but a herbicide treatment can be as good or better than any other method of site preparation.

Competition Indices

The study of competition dynamics requires both a measure of the competition level and a measure of crop response. Various competition indices have been developed, e.g., by Bella (1971) and Hegyi (1974) based on stem diameter, by Arney (1972), Ek and Monserud (1974), and Howard and Newton (1984) based on canopy development, and Daniels (1976), Wagner (1982), and Weiner (1984) with proximity-based models. Studies generally considered tree response to competition in terms of absolute height or basal area, but Zedaker (1982) and Brand (1986) sought to quantify crop tree size and environmental influences by using relative growth measures.

Tending

Tending is the term applied to pre-harvest silvicultural treatment of forest crop trees at any stage after initial planting or seeding. The treatment can be of the crop itself (e.g., spacing, pruning, thinning, and improvement cutting) or of competing vegetation (e.g., weeding, cleaning).

Planting

How many trees per unit area (spacing) that should be planted is not an easily answered question. Establishment density targets or regeneration standards have commonly been based on traditional practice, with the implicit aim of getting the stand quickly to the free-to-grow stage. Money is wasted if more trees are planted than are needed to achieve desired stocking rates, and the chance to establish other plantations is proportionately diminished. Ingress (natural regeneration) on a site is difficult to predict and often becomes surprisingly evident only some years after planting has been carried out. Early stand development after harvesting or other disturbance undoubtedly varies greatly among sites, each of which has its own peculiar characteristics.

For all practical purposes, the total volume produced by a stand on a given site is constant and optimum for a wide range of density or stocking. It can be decreased, but not increased, by altering the amount of growing stock to levels outside this range. Initial density affects stand development in that close spacing leads to full site utilization more quickly than wider spacing. Economic operability can be advanced by wide spacing even if total production is less than in closely spaced stands.

Beyond the establishment stage, the relationship of average tree size and stand density is very important. Various density-management diagrams conceptualizing the density-driven stand dynamics have been developed. Smith and Brand's (1988) diagram has mean tree volume on the vertical axis and the number of trees/ha on the horizontal axis: a stand can either have many little trees or a few big ones. The self-thinning line shows the largest number of trees of a given size/ha that can be carried at any given time. However, Willcocks and Bell (1995) caution against using such diagrams unless specific knowledge of the stand trajectory is known.

In the Lake States, plantations have been made with the spacing between trees varying from 3 by 3 to 10 by 10 feet (0.9 m by 0.9 m to 3.0 m by 3.0 m). Kittredge recommended that no fewer than 600 established trees per acre (1483/ha) be present during the early life of a plantation. To insure this, at least 800 trees per acre (1077/ha) should be planted where 85% survival may be expected, and at least 1200/ac (2970/ha) if only half of them can be expected to live. This translates into recommended spacings of 5 by 5 to 8 by 8 feet (1.5 m by 1.5 m to 2.4 m by 2.4 m) for plantings of conifers, including white spruce in the Lake States.

Enrichment Planting

A strategy for enhancing natural forests' economic value is to increase their concentration of economically important, indigenous tree species by planting seeds or seedlings for future harvest, which can be accomplished with enrichment planting (EP). This means increasing the planting density (i.e., the numbers of plants per hectare) in an already growing forest stand."

Release Treatments

- Weeding: A process of getting rid of saplings' or seedlings' competition by mowing, application of herbicide, or other method of removal from the surroundings.

- Cleaning: Release of select saplings from competition by overtopping trees of a comparable age. The treatment favors trees of a desired species and stem quality.

- Liberation cutting: A treatment that releases tree seedling or saplings by removing older overtopping trees.

Spacing

Over-crowded regeneration tends to stagnate. The problem is aggravated in species that have little self-pruning ability, such as white spruce. Spacing is a thinning (of natural regeneration), in which all trees other than those selected for retention at fixed intervals are cut. The term juvenile spacing is used when most or all of the cut trees are unmerchantable. Spacing can be used to obtain any of a wide range of forest management objectives, but it is especially undertaken to reduce density and control stocking in young stands and prevent stagnation, and to shorten the rotation, i.e., to speed the production of trees of a given size. Volume growth of individual trees and the merchantable growth of stands are increased. The primary rationale for spacing is that thinning is the projected decline in maximum allowable cut. And since wood will be concentrated on fewer, larger, and more uniform stems, operating and milling costs will be minimized.

Methods for spacing may be: manual, using various tools, including power saws, brush saws, and clippers; mechanical, using choppersand mulchers; chemical; or combinations of several methods. One treatment has had notable success in spacing massively overstocked (<100 000 stems/ha) natural regeneration of spruce and fir in Maine. Fitted to helicopter, the Thru-Valve boom emits herbicide spray droplets 1000 μm to 2000 μm in diameter at very low pressure. Swaths 1.2 m wide and leave strips 2.4 m wide were obtained with "knife-edge" precision when the herbicide was applied by helicopter flying at a height of 21 m at a speed of 40–48 km/h. It seems likely that no other method could be as cost-effective.

Twenty years after spacing to 2.5 × 2.5 m, 30-year-old mixed stands of balsam fir and white spruce in the Green River watershed, New Brunswick, averaged 156.9 m³/ha.

A spacing study of 3 conifers (white spruce, red pine and jack pine) was established at Moodie, Manitoba, on flat, sandy, nutritionally poor soils with a fresh moisture regime. Twenty years after planting, red pine had the largest average dbh, 15% greater than jack pine, while white spruce dbh was less than half that of the pines. Crown width showed a gradual increase with spacing for all 3 conifers. Results to date were suggesting optimum spacings between 1.8 m and 2.4 m for both pines; white spruce was not recommended for planting on such sites.

Comparable data are generated by espacement trials, in which trees are planted at a range of densities. Spacings of 1.25 m, 1.50 m, 1.75 m, 2.00 m, 2.50 m, and 3.00 m on 4 site classes were used in the 1922 trial at Petawawa, Ontario. In the first of 34 old field white spruce plantations used to investigate stand development in relation to spacing at Petawawa, Ontario, regular rows were planted at average spacings of from 4 × 4 to 7 × 7 feet (1.22 m × 1.22 m to 2.13 m × 2.13 m). Spacings up to 10 × 10 feet (3.05 m × 3.03 m) were subsequently included in the study. Yield tables based on 50 years of data showed:

a) Except for merchantable volumes at age 20 and site classes 50 and 60, closer spacings gave greater standing volumes at all ages than did wider spacings, the relative difference decreasing with age.

b) Merchantable volume as a proportion of total volume increases with age, and is greater at wider than at closer spacings.

c) Current annual volume increment culminates sooner at closer than at wider spacings.

A smaller espacement trial, begun in 1951 near Thunder Bay, Ontario, included white spruce at spacings of 1.8 m, 2.7 m, and 3.6 m. At the closest spacing, mortality had begun at 37 years, but not at the wider spacings.

The oldest interior spruce espacement trial in British Columbia was established in 1959 near Houston in the Prince Rupert Forest Region. Spacings of 1.2 m, 2.7 m, 3.7 m, and 4.9 m were used, and trees were measured 6, 12, 16, 26, and 30 years after planting. At wide espacements, trees developed larger diameters, crowns, and branches, but (at 30 years) basal area and total volume/ha were greatest in the closest espacement (Table 6.38). In more recent trials in the Prince George Region of British Columbia (Table 6.39) and in Manitoba, planting density of white spruce had no effect on growth after up to 16 growing seasons, even at spacings as low as 1.2 m. The slowness of juvenile growth and of crown closure delay the response to intra-competition. Initially, close spacing might even provide a positive nurse effect to offset any negative response to competition.

Thinning

Thinning is an operation that artificially reduces the number of trees growing in a stand with the aim of hastening the development of the remainder. The goal of thinning is to control the amount and distribution of available growing space. By altering stand density, foresters can influence the growth, quality, and health of residual trees. It also provides an opportunity to capture mortality and cull the commercially less desirable, usually smaller and malformed, trees. Unlike regeneration treatments, thinnings are not intended to establish a new tree crop or create permanent canopy openings.

Thinning greatly influences the ecology and micro-meteorology of the stand, lowering the inter-tree competition for water. The removal of any tree from a stand has repercussions on the remaining trees both above-ground and below. Silvicultural thinning is a powerful tool that can be used to influence stand development, stand stability, and the characteristics of the harvestable products.

When considering intensive conifer plantations designed for maximum production, it is essential to remember that tending and thinning regimes and wind and snow damage are intimately related.

Previous studies have demonstrated that repeated thinnings over the course of a forest rotation increase carbon stores relative to stands that are clear-cut on short rotations and that the carbon benefits differ according to thinning method (e.g., thinning from above versus below).

Precommercial Thinning

When natural regeneration or artificial seeding has resulted in dense, overstocked young stands, natural thinning will in most cases eventually reduce stocking to more silviculturally desirable levels. But by the time some trees reach merchantable size, others will be overmature and defective, and others will still be unmerchantable. The yield of merchantable wood can be greatly increased and the rotation shortened by pre-commercial thinning. Mechanical and chemical methods have been applied, but their costliness has militated against their ready adoption.

Pruning

Pruning, as a silvicultural practice, refers to the removal of the lower branches of the young trees (also giving the shape to the tree) so clear knot-free wood can subsequently grow over the branch stubs. Clear knot-free lumber has a higher value. Pruning has been extensively carried out in the Radiata pine plantations of New Zealand and Chile, however the development of Finger joint technology in the production of lumber and mouldings has led to many forestry companies reconsidering their pruning practices. "Brashing" is an alternative name for the same process. Pruning can be done to all trees, or more cost effectively to a limited number of trees. There are two types of pruning: natural or self-pruning and artificial pruning. Most cases of self-pruning happen when branches do not receive enough sunlight and die. Wind can also take part in natural pruning which can break branches. Artificial pruning is where people are paid to come and cut the branches. Or it can be natural, where trees are planted close enough that the effect is to cause self-pruning of low branches as energy is put into growing up for light reasons and not branchiness.

Stand Conversion

The term *stand conversion* refers to a change from one silvicultural system to another and includes *species conversion*, i.e., a change from one species (or set of species) to another. Such change can be effected intentionally by various silvicultural means, or incidentally by default e.g., when high-grading has removed the coniferous content from a mixedwood stand, which then becomes exclusively self-perpetuating aspen. In general, such sites as these are the most likely to be considered for conversion.

Growth and Yield

In discussing yields that might be expected from the Canadian spruce forests, Haddock (1961) noted that Wright's (1959) quotation of spruce yields in the British Isles of 220 cubic feet per acre (15.4 m³/ha) per year and in Germany of 175 cubic feet per acre (12.25 m³/ha) per year was misleading, at least if it was meant to imply that such yields might be approached in the Boreal Forest Region of Canada. Haddock thought that Wright's suggestion of 20 to 40 (average 30) cubic feet per acre (1.4 m³/ha to 2.8 m³/ha (average 2.1 m³/ha) per year was more reasonable, but still somewhat optimistic.

The principal way forest resource managers influence growth and yield is to manipulate the mixture of species and number (density) and distribution (stocking) of individuals that form the canopy of the stand. Species composition of much of the boreal forest in North America already differs greatly from its pre-exploitation state. There is less spruce and more hardwoods in the second-growth forest than in the original forest; Hearnden et al. (1996) calculated that the spruce cover type had declined from 18% to only 4% of the total forested area in Ontario. Mixedwood occupies a greater proportion of Ontario's second-growth forest (41%) than in the original (36%), but its component of white spruce is certainly much diminished.

Growth performance is certainly influenced by site conditions and thus by the kind and degree of site preparation in relation to the nature of the site. It is important to avoid the assumption that site preparation of a particular designation will have a particular silvicultural outcome. Scarification, for instance, not only covers a wide range of operations that scarify, but also any given way of scarifying can have significantly different results depending on site conditions at the time of treatment. In point of fact, the term is commonly misapplied. *Scarification* is defined as "Loosening the top soil of open areas, or breaking up the forest floor, in preparation for regenerating by direct seeding or natural seedfall", but the term is often misapplied to practices that include scalping, screefing, and blading, which pare off low and surface vegetation, together with most off its roots to expose a weed-free surface, generally in preparation for sowing or planting thereon.

Thus, it is not surprising that literature can be used to support the view that the growth of seedlings on scarified sites is much superior to that of growth on similar sites that have not been scarified, while other evidence supports the contrary view that scarification can reduce growth. Detrimental results can be expected from scarification that impoverishes the rooting zone or exacerbates edaphic or climatic constraints.

Burning site preparation has enhanced spruce seedling growth, but it must be supposed that burning could be detrimental if the nutrient capital is significantly depleted.

An obvious factor greatly influencing regeneration is competition from other vegetation. In a pure stand of Norway spruce, for instance, Roussel (1948) found the following relationships:

Percent cover (%)	Vegetation Description
Below 1	No vegetation
1-3	Moss carpet with a few fir seedlings
4-10	Herbaceous plants appear
10-25	Bramble, herbs, fairly vigorous spruce seedlings
>25	Herbs, brambles very dense, vigorous, no moss

A factor of some importance in solar radiation–reproduction relationships is excess heating of the soil surface by radiation. This is especially important for seedlings, such as spruce, whose first leaves do not shade the base of the stem at the soil surface. Surface temperatures in sandy soils on occasion reach lethal temperatures of 50 °C to 60 °C.

Common Methods of Harvesting

Silvicultural regeneration methods combine both the harvest of the timber on the stand and re-establishment of the forest. The proper practice of sustainable forestry should mitigate the potential negative impacts, but all harvest methods will have some impacts on the land and residual stand. The practice of sustainable forestry limits the impacts such that the values of the forest are maintained in perpetuity. Silvicultural prescriptions are specific solutions to a specific set of circumstances and management objectives. Following are some common methods:

Clearcut Harvesting

Conventional clearcut harvesting is relatively simple: all trees on a cutblock are felled and bunched with bunches aligned to the skidding direction, and a skidder then drags the bunches to the closest log deck. Feller-buncher operators concentrate on the width of the felled swath, the number of trees in a bunch, and the alignment of the bunch. Providing a perimeter boundary is felled during daylight, night-shift operations can continue without the danger of trespassing beyond the block. Productivity of equipment is maximized because units can work independently of one another.

Clearcutting

An even-aged regeneration method that can employ either natural or artificial regeneration. It involves the complete removal of the forest stand at one time. Clearcutting can be biologically appropriate with species that typically regenerate from stand replacing fires or other major disturbances, such as Lodgepole Pine (*Pinus contorta*). Alternatively, clearcutting can change the dominating species on a stand with the introduction of non-native and invasive species as was shown at the Blodgett Experimental Forest near Georgetown California. Additionally, clearcutting can prolong slash decomposition, expose soil to erosion, impact visual appeal of a landscape and remove essential wildlife habitat. It is particularly useful in regeneration of tree species such as Douglas-fir (*Pseudotsuga menziesii*) which is shade intolerant. In addition, the general public's distaste for even-aged silviculture, particularly clearcutting, is likely to result in a greater role for uneven-aged management on public lands as well. Across Europe, and in parts of North America, even-aged, production-orientated and intensively managed plantations are beginning to be regarded in the same way as old industrial complexes: something to abolish or convert to something else.

Clearcutting will impact many site factors important in their effect on regeneration, including air and soil temperatures. Kubin and Kemppainen (1991), for instance, measured temperatures in northern Finland from 1974 through 1985 in 3 clear-felled areas and in 3 neighouring forest stands dominated by Norway spruce. Clear felling had no significant influence on air temperature at 2 m above the ground surface, but the daily air temperature maxima at 10 cm were greater in the clear-felled area than in the uncut forest, while the daily minima at 10 cm were lower. Night frosts were more common in the clear-felled area. Daily soil temperatures at 5 cm depth were 2 °C to 3 °C greater in the clear-felled area than in the uncut forest, and temperatures at depths of 50 cm and 100 cm were 3 °C to 5 °C greater. The differences between the clear-felled and uncut areas did not diminish during the 12 years following cutting.

Coppicing

A regeneration method which depends on the sprouting of cut trees. Most hardwoods, the coast redwood, and certain pines naturally sprout from stumps and can be managed through coppicing. Coppicing is generally used to produce fuelwood, pulpwood, and other products dependent on small trees. A close relative of coppicing is pollarding. Three systems of coppice woodland management are generally recognized: simple coppice, coppice with standards, and the coppice selection system.

- In Compound coppicing or coppicing with standards, some of the highest quality trees are retained for multiple rotations in order to obtain larger trees for different purposes.

Direct Seeding

Prochnau (1963), 4 years after sowing, found that 14% of viable white spruce seed sown on mineral soil had produced surviving seedlings, at a seed:seedling ratio of 7.1:1. With Engelmann spruce, Smith and Clark (1960) obtained average 7th year seed:seedling ratios of 21:1 on scarified seedbeds on dry sites, 38:1 on moist sites, and 111:1 on litter seedbeds.

Group Selection

The group selection method is an uneven-aged regeneration method that can be used when mid-tolerant species regeneration is desired. The group selection method can still result in residual stand damage in dense stands, however directional falling can minimize the damage. Additionally, foresters can select across the range of diameter classes in the stand and maintain a mosaic of age and diameter classes.

Méthode Du Contrôle

Classical European silviculture achieved impressive results with systems such as Henri Biolley's *méthode du contrôle* in Switzerland, in which the number and size of trees harvested were determined by reference to data collected from every tree in every stand measured every 7 years.

While not designed to be applied to boreal mixedwoods, the *méthode du contrôle* is described briefly here to illustrate the degree of sophistication applied by some European foresters to the management of their forests. Development of management techniques that allowed for stand development to be monitored and guided into sustainable paths were in part a response to past experience, particularly in Central European countries, of the negative effects of pure, uniform stands with species often unsuited to the site, which greatly increased the risk of soil degradation and biotic diseases. Increased mortality and decreased increment generated widespread concern, especially after reinforcement by other environmental stresses.

More or less uneven-aged, mixed forests of preponderantly native species, on the other hand, treated along natural lines, have proved to be healthier and more resistant to all kinds of external dangers; and in the long run such stands are more productive and easier to protect.

However, irregular stands of this type are definitely more difficult to manage—new methods and techniques had to be sought particularly for the establishment of inventories, as well as increment control and yield regulation. In Germany, for instance, since the beginning of the nineteenth century under the influence of G.L. Hartig (1764–1837), yield regulation has been effected almost exclusively by allotment or formula methods based on the conception of the uniform normal forest with a regular succession of cutting areas.

In France, on the other hand, efforts were made to apply another kind of forest management, one that aimed to bring all parts of the forest to a state of highest productive capacity in perpetuity. In 1878, the French forester A. Gurnaud (1825–1898) published a description of a *méthode du contrôle* for determining increment and yield. The method was based on the fact that through careful, selective harvesting, the productivity of the residual stand can be improved, because timber is removed as a cultural operation. In this method, the increment of stands is accurately determined periodically with the object of gradually converting the forest, through selective management and continuous experimentation, to a condition of equilibrium at maximum productive capacity.

Henri Biolley (1858–1939) was the first to apply Gurnaud's inspired ideas to practical forestry. From 1890 on, he managed the forests of his Swiss district according to these principles, devoting himself for almost 50 years to the study of increment and a treatment of stands directed towards the highest production, and proving the practicability of the check method. In 1920, he published this study giving a theoretical basis of management of forests under the check method, describing the procedures to be applied in practice (which he partly developed and simplified), and evaluating the results.

Biolley's pioneering work formed the basis upon which most Swiss forest management practices were later developed, and his ideas have been generally accepted. Today, with the trend of intensifying forest management and productivity in most countries, the ideas and application of careful, continuous treatment of stands with the aid of the volume check method are meeting with ever-growing interest. In Britain and Ireland, for example, there is increased application of Continuous Cover Forestry principles to create permanently irregular structures in many woodlands.

Patch Cut

Row and Broadcast Seeding

Spot and row seeders use less seed that does broadcast ground or aerial seeding but may induce clumping. Row and spot seeding confer greater ability to control seed placement than does broadcast seeding. Also, only a small percentage of the total area needs to be treated.

In the aspen type of the Great Lakes region, direct sowing of the seed of conifers has usually failed. However, Gardner (1980) after trials in Yukon, which included broadcast seeding of white spruce seed at 2.24 kg/ha that secured 66.5% stocking in the Scarified Spring Broadcast treatment 3 years after seeding, concluded that the technique held "considerable promise".

Seed-tree

An even-aged regeneration method that retains widely spaced residual trees in order to provide uniform seed dispersal across a harvested area. In the seed-tree method, 2-12 seed trees per acre (5-30/ha) are left standing in order to regenerate the forest. They will be retained until regeneration has become established at which point they may be removed. It may not always be economically viable or biologically desirable to re-enter the stand to remove the remaining seed trees. Seed-tree cuts can also be viewed as a clearcut with natural regeneration and can also have all of the problems associated with clearcutting. This method is most suited for light-seeded species and those not prone to windthrow.

Selection Systems

Selection systems are appropriate where uneven stand structure is desired, particularly where the need to retain continuous cover forest for aesthetic or environmental reasons outweighs other management considerations. Selection logging has been suggested as being of greater utility than shelterwood systems in regenerating old-growth Engelmann Spruce Sub-alpine Fir (ESSF) stands in southern British Columbia. In most areas, selection logging favours regeneration of fir more than the more light-demanding spruce. In some areas, selection logging can be expected to favour spruce over less tolerant hardwood species (Zasada 1972) or lodgepole pine.

Shelter Spot Seeding

The use of shelters to improve germination and survival in spot seedings seeks to capture the benefits of greenhouse culture, albeit miniature. The Hakmet seed shelter, for instance, is a semi-transparent plastic cone 8 cm high, with openings of 7 cm diameter in the 7.5 cm diameter base and 17 mm diameter in the 24 mm diameter top. This miniature greenhouse increases air humidity, reduces soil desiccation, and raises air and soil temperatures to levels more favourable to germination and seedling growth than those offered by unprotected conditions. The shelter is designed to break down after a few years of exposure to ultraviolet radiation.

Seed shelters and spring sowing significantly improved stocking compared with bare spot seeding, but sheltering did not significantly improve growth. Stocking of bare seedspots was extremely low, possibly due to smothering of seedlings by abundant broadleaf and herbaceous litter, particularly that from aspen and red raspberry, and exacerbated by strong competition from graminoids and raspberry.

Cone shelters (Cerkon™) usually produced greater survival than unsheltered seeding on scarified seedspots in trials of direct seeding techniques in interior Alaska, and funnel shelters (Cerbel™) usually produced greater survival than unsheltered seeding on non-scarified seedspots. Both shelter types are manufactured by AB Cerbo in Trollhättan, Sweden. Both are made of light-degradable, white, opaque plastic, and are 8 cm high when installed.

White spruce seed was sown in Alaska on a burned site in summer 1984, and protected by white plastic cones on small spots scarified by hand, or by white funnels placed directly into the residual ash and organic material. A group of 6 ravens (*Corvus corax*) was observed in the area about 1 week after sowing was completed in mid-June. Damage averaged 68% with cones and 50% with funnels on an upland area, and 26% with funnels on a floodplain area. Damage by ravens was only 0.13% on unburned but otherwise similar areas.

In seeding trials in Manitoba between 1960 and 1966 aimed at converting aspen stands to spruce–aspen mixedwoods, 1961 scarification in the Duck Mountain Provincial Forest remained receptive to natural seeding for many years.

Shelterwood

In general terms, the shelterwood system is a series of partial cuts that removes the trees of an existing stand over several years and eventually culminates in a final cut that creates a new even-aged stand. It is an even-aged regeneration method that removes trees in a series of three harvests: 1) Preparatory cut; 2) Establishment cut; and 3) Removal cut. The success of practising a shelterwood system is closely related to: 1. the length of the regeneration period, i.e. the time from the shelterwood cutting to the date when a new generation of trees has been established; 2.the quality of the new tree stand with respect to stand density and growth; and 3.the value increment of the shelter trees. Information on the establishment, survival and growth of seedlings influenced by the cover of shelter trees, as well as on the growth of these trees, is needed as a basis for modelling the economic return of practising a shelterwood system. The method's objective is to establish new forest reproduction under the shelter of the retained trees. Unlike the seed-tree method, residual trees alter understory environmental conditions (i.e. sunlight, temperature, and moisture) that influence tree seedling growth. This method can also find a middle ground with the light ambiance by having less light accessible to competitors while still being able to provide enough light for tree regeneration. Hence, shelterwood methods are most often chosen for site types characterized by extreme conditions, in order to create a new tree generation within a reasonable time period. These conditions are valid foremost on level ground sites which are either dry and poor or moist and fertile.

Shelterwood Systems

Shelterwood systems involve 2, 3, or exceptionally more partial cuttings. A final cut is made once adequate natural regeneration has been obtained. The shelterwood system is most commonly applied as a 2-cut uniform shelterwood, first an initial regeneration (seed) cut, the second a final harvest cut. In stands less than 100 years old, a light preparatory cut can be useful. A series of intermediate cuts at intervals of 10–20 years has been recommended for intensively managed stands.

From operational or economic standpoints, however, there are disadvantages to the shelterwood system: harvesting costs are higher; trees left for deferred cutting may be damaged during the regeneration cut or related extraction operations; the increased risk of blowdown threatens the seed source; damage from bark beetles is likely to increase; regeneration may be damaged during the final cut and related extraction operations; the difficulty of any site preparation would be increased; and incidental damage to regeneration might be caused by any site preparation operations.

Single-tree Selection

The single-tree selection method is an uneven-aged regeneration method most suitable when shade tolerant species regeneration is desired. It is typical for older and diseased trees to be removed, thus thinning the stand and allowing for younger, healthy trees to grow. Single-tree selection can be very difficult to implement in dense or sensitive stands and residual stand damage can occur. This method is also disturbs the canopy layer the least out of all other methods.

Spot Seeding

Spot seeding was found to be the most economical and reliable of the direct seeding methods for converting aspen and paper birch to spruce and pine. In the Chippewa National Forest (Lake States), seed-spot sowing of 10 seeds each of white spruce and white pine under 40-year aspen after different degrees of cutting on gave second-season results clearly indicating the need to remove or disturb the forest floor to obtain germination of seeded white spruce and white pine.

Spot seeding of coniferous seed, including white spruce, has had occasional success, but several constraining factors commonly limit germination success: the drying out of the forest floor before the roots of germinants reach underlying moisture reserves; and, particularly under hardwoods, the smothering of small seedlings by snow-pressed leaf litter and lesser vegetation. Kittredge and Gervorkiantz (1929) determined that removal of the aspen forest floor increased germination percentage after the second season in seed spots of both white pine and white spruce, in 4 plots, from 2.5% to 5%, from 8% to 22%, from 1% to 9.5%, and from 0% to 15%.

Spot seeding requires less seed than broadcast seeding and tends to achieve more uniform spacing, albeit sometimes with clumping. The devices used in Ontario for manual spot seeding are the "oil can" seeder, seeding sticks, and shakers. The oil can is a container fitted with a long spout through which a predetermined number of seeds are released with each flick of the seeder.

Strip Cutting

Harvesting cutblocks where only a portion of the trees are to be removed is very different from clearcutting. First, trails must be located to provide access for the felling and skidding/forwarding equipment. These trails must be carefully located to ensure that the trees remaining meet the desired quality criteria and stocking density. Second, the equipment must not damage the residual stand. The further desiderata are outlined by Sauder (1995).

The dearth of seed and a deficiency of receptive seedbeds were recognized as major reasons for the lack of success of clearcut harvesting. One remedy attempted in British Columbia and Alberta has been alternate strip cutting. The greater seed source from uncut trees between the cut strips, and the disturbance to the forest floor within the cut strips could be expected to increase the amount of natural regeneration. Trees were cut to a diameter limit in the cut strips, but large trees in the leave strips often proved too much of a temptation and were cut too, thus removing those trees that would otherwise have been the major source of seed.

An unfortunate consequence of strip thinning was the build-up of spruce beetle populations. Shaded slash from the initial cut, together with an increase in the number of windthrown trees in the leave strips, provided conditions ideally suited to the beetle.

Underplanting

DeLong et al. (1991) suggested underplanting 30- to 40-year-old aspen stands, on the basis of the success of natural spruce in regenerating under stands of such stands: "By planting, spacing can be controlled enabling easier protection of the spruce during stand entry for harvesting of the aspen overstorey".

Variable Retention

A harvesting and regeneration method which is a relatively new silvicultural system that retains forest structural elements (stumps, logs, snags, trees, understory spieces and undisturbed layers of forest floor) for at least one rotation in order to preserve environmental values associated with structurally complex forests.

"Uneven-aged and even-aged methods differ in the scale and intensity of disturbance. Uneven-aged methods maintain a mix of tree sizes or ages within a habitat patch by periodically harvesting individual or small groups of trees, Even-aged methods harvest most or all of the overstory and create a fairly uniform habitat patch dominated by trees of the same age". Even-aged management systems have been the prime methods to use when studying the effects on birds.

Mortality

A survey in 1955–56 to determine survival, development, and the reasons for success or failure of conifer pulpwood plantations (mainly of white spruce) in Ontario and Quebec up to 32 years old found that the bulk of the mortality occurred within the first 4 years of planting, unfavourable site and climate being the main causes of failure.

Advance Growth

Naturally regenerated trees in an understorey prior to harvesting constitute a classic case of good news and bad news. Understorey white spruce is of particular importance in mixedwoods dominated by aspen, as in the B15, B18a, and B19a Sections of Manitoba, and elsewhere. Until the latter part of the last century, white spruce understorey was mostly viewed as money in the bank on a long-term, low interest deposit, with final yield to be realized after slow natural succession, but the resource became increasingly threatened with the intensification of harvesting of aspen. White spruce plantations on mixedwood sites proved expensive, risky, and generally unsuccessful. This prompted efforts to see what might be done about growing aspen and white spruce on the same landbase by protecting existing white spruce advance growth, leaving a range of viable crop trees during the first cut, then harvesting both hardwoods and spruce in the final cut. Information about the understorey component is critical to spruce management planning. The ability of then current harvesting technology and crews employed to provide adequate protection for white spruce understories was questioned by Brace and Bella. Specialized equipment and training, perhaps with financial incentives, may be needed to develop procedures that would confer the degree of protection needed for the system to be feasible. Effective understorey management planning requires more than improved mixedwood inventory.

Avoidance of damage to the understorey will always be a desideratum. Sauder's (1990) paper on mixedwood harvesting describes studies designed to evaluate methods of reducing non-trivial damage to understorey residuals that would compromise their chance of becoming a future crop tree. Sauder concluded that: (1) operational measures that protected residual stems may not unduly increase costs, (2) all felling, conifers and hardwoods, needs to be done in one operation to minimize the entry of the feller-buncher into the residual stand, (3) several operational procedures can reduce understorey damage, some of them without incurring extra costs, and (4) successful harvesting of treatment blocks depends primarily on the intelligent location of skid trails and landings. In summary, the key to protecting the white spruce understorey without sacrificing logging efficiency is a combination of good planning, good supervision, the use of appropriate equipment, and having conscientious, well-trained operators.Even the best plan will not reduce understorey damage unless its implementation is supervised.

New stands need to be established to provide for future supply of commercial white spruce from 150 000 ha of boreal mixedwoods in 4 of Rowe's (1972) regional Forest Sections straddling Alberta, Saskatchewan, and Manitoba, roughly from Peace River AB to Brandon MB. In the 1980s, with harvesting using conventional equipment and procedures, a dramatic increase in the demand for aspen posed a serious problem for the associated spruce understorey. Formerly, white spruce in the understories had developed to commercial size through natural succession under the protection of the hardwoods. Brace articulated a widespread concern: "The need for protection of spruce as a component of boreal mixedwoods goes beyond concern for the future commercial softwood timber supply. Concerns also include fisheries and wildlife habitat, aesthetics and recreation, a general dissatisfaction with cleacutting in mixedwoods and a strong interest in mixedwood perpetuation, as expressed recently in 41 public meetings on forestry development in northern Alberta...".

On the basis of tests of 3 logging systems in Alberta, Brace (1990) affirmed that significant amounts of understorey can be retained using any of those systems provided that sufficient effort is directed towards protection. Potential benefits would include increased short-term softwood timber supply, improved wildlife habitat and cutblock aesthetics, as well as reduced public criticism of previous logging practices. Stewart et al. (2001) developed statistical models to predict the natural establishment and height growth of understorey white spruce in the boreal mixedwood forest in Alberta using data from 148 permanent sample plots and supplementary information about height growth of white spruce regeneration and the amount and type of available substrate. A discriminant model correctly classified 73% of the sites as to presence or absence of a white spruce understorey, based on the amount of spruce basal area, rotten wood, ecological nutrient regime, soil clay fraction, and elevation, although it explained only 30% of the variation in the data. On sites with a white spruce understorey, a regression model related the abundance of regeneration to rotten wood cover, spruce basal area, pine basal area, soil clay fraction, and grass cover ($R^2 = 0.36$). About half the seedlings surveyed grew on rotten wood, and only 3% on mineral soil, and seedlings were 10 times more likely to have established on these substrates than on litter. Exposed mineral soil covered only 0.3% of the observed transect area.

Advance Growth Management

Advance growth management, i.e., the use of suppressed understorey trees, can reduce reforestation costs, shorten rotations, avoid denuding the site of trees, and also reduce adverse impacts on aesthetic, wildlife, and watershed values. To be of value, advance growth must have acceptable species composition and distribution, have potential for growth following release, and not be vulnerable to excessive damage from logging.

The age of advance growth is difficult to estimate from its size, as white that appears to be 2- to 3-year-old may well be more than 20 years old. However, age does not seem to determine the ability of advance growth of spruce to respond to release, and trees older than 100 years have shown rapid rates of growth after release. Nor is there a clear relationship between the size of advance growth and its growth rate when released.

Where advance growth consists of both spruce and fir, the latter is apt to respond to release more quickly than the former, whereas spruce does respond. If the ratio of fir to spruce is large, however, the greater responsiveness to release of fir may subject the spruce to competition severe enough to negate much of the effect of release treatment. Even temporary relief from shrub competition has increased height growth rates of white spruce in northwestern New Brunswick, enabling the spruce to overtop the shrubs.

Site Preparation

Site preparation is any of various treatments applied to a site in order to ready it for seeding or planting. The purpose is to facilitate the regeneration of that site by the chosen method. Site preparation may be designed to achieve, singly or in any combination: improved access, by reducing or rearranging slash, and amelioration of adverse forest floor, soil, vegetation, or other biotic factors. Site preparation is undertaken to ameliorate one or more constraints that would otherwise be likely to thwart the objectives of management. A valuable bibliography on the effects of soil temperature and site preparation on subalpine and boreal tree species has been prepared by McKinnon et al. (2002).

Site preparation is the work that is done before a forest area is regenerated. Some types of site preparation are burning.

Burning

Broadcast burning is commonly used to prepare clearcut sites for planting, e.g., in central British Columbia, and in the temperate region of North America generally.

Prescribed burning is carried out primarily for slash hazard reduction and to improve site conditions for regeneration; all or some of the following benefits may accrue:

a)   Reduction of logging slash, plant competition, and humus prior to direct seeding, planting, scarifying or in anticipation of natural seeding in partially cut stands or in connection with seed-tree systems.

b)   Reduction or elimination of unwanted forest cover prior to planting or seeding, or prior to preliminary scarification thereto.

c)   Reduction of humus on cold, moist sites to favour regeneration.

d)   Reduction or elimination of slash, grass, or brush fuels from strategic areas around forested land to reduce the chances of damage by wildfire.

Prescribed burning for preparing sites for direct seeding was tried on a few occasions in Ontario, but none of the burns was hot enough to produce a seedbed that was adequate without supplementary mechanical site preparation.

Changes in soil chemical properties associated with burning include significantly increased pH, which Macadam (1987) in the Sub-boreal Spruce Zone of central British Columbia found persisting more than a year after the burn. Average fuel consumption was 20 to 24 t/ha and the forest floor depth was reduced by 28% to 36%. The increases correlated well with the amounts of slash (both total and ≥7 cm diameter) consumed. The change in pH depends on the severity of the burn and the amount consumed; the increase can be as much as 2 units, a 100-fold change. Deficiencies of copper and iron in the foliage of white spruce on burned clearcuts in central British Columbia might be attributable to elevated pH levels.

Even a broadcast slash fire in a clearcut does not give a uniform burn over the whole area. Tarrant (1954), for instance, found only 4% of a 140-ha slash burn had burned severely, 47% had burned lightly, and 49% was unburned. Burning after windrowing obviously accentuates the subsequent heterogeneity.

Marked increases in exchangeable calcium also correlated with the amount of slash at least 7 cm in diameter consumed. Phosphorus availability also increased, both in the forest floor and in the 0 cm to 15 cm mineral soil layer, and the increase was still evident, albeit somewhat diminished, 21 months after burning. However, in another study in the same Sub-boreal Spruce Zone found that although it increased immediately after the burn, phosphorus availability had dropped to below pre-burn levels within 9 months.

Nitrogen will be lost from the site by burning, though concentrations in remaining forest floor were found by Macadam (1987) to have increased in 2 of 6 plots, the others showing decreases. Nutrient losses may be outweighed, at least in the short term, by improved soil microclimate through the reduced thickness of forest floor where low soil temperatures are a limiting factor.

The *Picea/Abies* forests of the Alberta foothills are often characterized by deep accumulations of organic matter on the soil surface and cold soil temperatures, both of which make reforestation difficult and result in a general deterioration in site productivity; Endean and Johnstone (1974) describe experiments to test prescribed burning as a means of seedbed preparation and site amelioration on representative clear-felled *Picea/Abies* areas. Results showed that, in general, prescribed burning did not reduce organic layers satisfactorily, nor did it increase soil temperature, on the sites tested. Increases in seedling establishment, survival, and growth on the burned sites were probably the result of slight reductions in the depth of the organic layer, minor increases in soil temperature, and marked improvements in the efficiency of the planting crews. Results also suggested that the process of site deterioration has not been reversed by the burning treatments applied.

Ameliorative Intervention

Slash weight (the oven-dry weight of the entire crown and that portion of the stem < 4 inches in diameter) and size distribution are major factors influencing the forest fire hazard on harvested sites. Forest managers interested in the application of prescribed burning for hazard reduction and silviculture, were shown a method for quantifying the slash load by Kiil (1968). In west-central Alberta, he felled, measured, and weighed 60 white spruce, graphed (a) slash weight per merchantable unit volume against diameter at breast height (dbh), and (b) weight of fine slash (<1.27 cm) also against dbh, and produced a table of slash weight and size distribution on one acre of a hypothetical stand of white spruce. When the diameter distribution of a stand is unknown, an estimate of slash weight and size distribution can be obtained from average stand diameter, number of trees per unit area, and merchantable cubic foot volume. The sample trees in Kiil's study had full symmetrical crowns. Densely growing trees with short and often irregular crowns would probably be overestimated; open-grown trees with long crowns would probably be underestimated.

The need to provide shade for young outplants of Engelmann spruce in the high Rocky Mountains is emphasized by the U.S. Forest Service. Acceptable planting spots are defined as microsites on the north and east sides of down logs, stumps, or slash, and lying in the shadow cast by such material. Where the objectives of management specify more uniform spacing, or higher densities, than obtainable from an existing distribution of shade-providing material, redistribution or importing of such material has been undertaken.

Access

Site preparation on some sites might be done simply to facilitate access by planters, or to improve access and increase the number or distribution of microsites suitable for planting or seeding.

Wang et al. (2000) determined field performance of white and black spruces 8 and 9 years after outplanting on boreal mixedwood sites following site preparation (Donaren disc trenching versus no trenching) in 2 plantation types (open versus sheltered) in southeastern Manitoba. Donaren trenching slightly reduced the mortality of black spruce but significantly increased the mortality of white spruce. Significant difference in height was found between open and sheltered plantations for black spruce but not for white spruce, and root collar diameter in sheltered plantations was significantly larger than in open plantations for black spruce but not for white spruce. Black spruce open plantation had significantly smaller volume (97 cm^3) compared with black spruce sheltered (210 cm^3), as well as white spruce open (175 cm^3) and sheltered (229 cm^3) plantations. White spruce open plantations also had smaller volume than white spruce sheltered plantations. For transplant stock, strip plantations had a significantly higher volume (329 cm^3) than open plantations (204 cm^3). Wang et al. (2000) recommended that sheltered plantation site preparation should be used.

Mechanical

Up to 1970, no "sophisticated" site preparation equipment had become operational in Ontario, but the need for more efficacious and versatile equipment was increasingly recognized. By this time, improvements were being made to equipment originally developed by field staff, and field testing of equipment from other sources was increasing.

According to J. Hall (1970), in Ontario at least, the most widely used site preparation technique was post-harvest mechanical scarification by equipment front-mounted on a bulldozer (blade, rake, V-plow, or teeth), or dragged behind a tractor (Imsett or S.F.I. scarifier, or rolling chopper). Drag type units designed and constructed by Ontario's Department of Lands and Forests used anchor chain or tractor pads separately or in combination, or were finned steel drums or barrels of various sizes and used in sets alone or combined with tractor pad or anchor chain units.

J. Hall's (1970) report on the state of site preparation in Ontario noted that blades and rakes were found to be well suited to post-cut scarification in tolerant hardwood stands for natural regeneration of yellow birch. Plows were most effective for treating dense brush prior to planting, often in conjunction with a planting machine. Scarifying teeth, e.g., Young's teeth, were sometimes used to prepare sites for planting, but their most effective use was found to be preparing sites for seeding, particularly in backlog areas carrying light brush and dense herbaceous growth. Rolling choppers found application in treating heavy brush but could be used only on stone-free soils. Finned drums were commonly used on jack pine–spruce cutovers on fresh brushy sites with a deep duff layer and heavy slash, and they needed to be teamed with a tractor pad unit to secure good distribution of the slash. The S.F.I. scarifier, after strengthening, had been "quite successful" for 2 years, promising trials were under way with the cone scarifier and barrel ring scarifier, and development had begun on a new flail scarifier for use on sites with shallow, rocky soils. Recognition of the need to become more effective and efficient in site preparation led the Ontario Department of Lands and Forests to adopt the policy of seeking and obtaining for field testing new equipment from Scandinavia and elsewhere that seemed to hold promise for Ontario conditions, primarily in the north. Thus, testing was begun of the Brackekultivator from Sweden and the Vako-Visko rotary furrower from Finland.

Mounding

Site preparation treatments that create raised planting spots have commonly improved outplant performance on sites subject to low soil temperature and excess soil moisture. Mounding can certainly have a big influence on soil temperature. Draper et al. (1985), for instance, documented this as well as the effect it had on root growth of outplants (Table 30).

The mounds warmed up quickest, and at soil depths of 0.5 cm and 10 cm averaged 10 and 7 °C higher, respectively, than in the control. On sunny days, daytime surface temperature maxima on the mound and organic mat reached 25 °C to 60 °C, depending on soil wetness and shading. Mounds reached mean soil temperatures of 10 °C at 10 cm depth 5 days after planting, but the control did not reach that temperature until 58 days after planting. During the first growing season, mounds had 3 times as many days with a mean soil temperature greater than 10 °C than did the control microsites.

Draper et al.'s (1985) mounds received 5 times the amount of photosynthetically active radiation (PAR) summed over all sampled microsites throughout the first growing season; the control treatment consistently received about 14% of daily background PAR, while mounds received over 70%. By November, fall frosts had reduced shading, eliminating the differential. Quite apart from its effect on temperature, incident radiation is also important photosynthetically. The average control microsite was exposed to levels of light above the compensation point for only 3 hours, i.e., one-quarter of the daily light period, whereas mounds received light above the compensation point for 11 hours, i.e., 86% of the same daily period. Assuming that incident light in the 100-600 $\mu Em^{-2}s^{-1}$ intensity range is the most important for photosynthesis, the mounds received over 4 times the total daily light energy that reached the control microsites.

Orientation of Linear Site Preparation, e.g., Disk-trenching

With linear site preparation, orientation is sometimes dictated by topography or other considerations, but the orientation can often be chosen. It can make a difference. A disk-trenching experiment in the Sub-boreal Spruce Zone in interior British Columbia investigated the effect on growth of young outplants (lodgepole pine) in 13 microsite planting positions: berm, hinge, and trench in each of north, south, east, and west aspects, as well as in untreated locations between the furrows. Tenth-year stem volumes of trees on south, east, and west-facing microsites were significantly greater than those of trees on north-facing and untreated microsites. However, planting spot selection was seen to be more important overall than trench orientation.

In a Minnesota study, the N–S strips accumulated more snow but snow melted faster than on E–W strips in the first year after felling. Snow-melt was faster on strips near the centre of the strip-felled area than on border strips adjoining the intact stand. The strips, 50 feet (15.24 m) wide, alternating with uncut strips 16 feet (4.88 m) wide, were felled in a *Pinus resinosa* stand, aged 90 to 100 years.

References

- Chantrell, Glynnis, ed. (2002). The Oxford Dictionary of Word Histories. Oxford University Press. p. 14. ISBN 0-19-863121-9.

- Committee on Forestry Research, National Research Council (1990). Forestry Research: A Mandate for Change. National Academies Press. pp. 15–16. ISBN 0-309-04248-8.

- Ensminger, M.E.; Parker, R.O. (1986). Sheep and Goat Science (Fifth ed.). Interstate Printers and Publishers. ISBN 0-8134-2464-X.

- Broudy, Eric (1979). The Book of Looms: A History of the Handloom from Ancient Times to the Present. UPNE. p. 81. ISBN 978-0-87451-649-4.

- National Geographic (2015). Food Journeys of a Lifetime. National Geographic Society. p. 126. ISBN 978-1-4262-1609-1.

- "World oil supplies are set to run out faster than expected, warn scientists". The Independent. 14 June 2007. Archived from the original on 21 October 2010. Retrieved 14 July 2016.

- "Africa may be able to feed only 25% of its population by 2025". Mongabay. 14 December 2006. Archived from the original on 27 November 2011. Retrieved 15 July 2016.

- "National Agroforestry Center". USDA National Agroforestry Center (NAC). Archived from the original on 19 August 2015. Retrieved 2 April 2014.

- "Silvopasture". Agroforestry Research Trust [in England]. Archived from the original on 20 April 2015. Retrieved 19 August 2015.

- International Food Policy Research Institute (2014). "Food Security in a World of Growing Natural Resource Scarcity". CropLife International. Retrieved 1 July 2013.

- Whiteside, Stephanie (28 November 2012). "Peru bans genetically modified foods as US lags". Current TV. Archived from the original on 24 March 2013. Retrieved 7 May 2013.

- "Livestock a major threat to environment". UN Food and Agriculture Organization. 29 November 2006. Archived from the original on 28 March 2008. Retrieved 24 April 2013.

- "Water Management: Towards 2030". FAO. March 2003. Archived from the original on 10 May 2013. Retrieved 7 May 2013.

- Runge, C. Ford (June 2006). "Agricultural Economics: A Brief Intellectual History" (PDF). Center for International Food and Agriculture Policy. p. 4. Retrieved 16 September 2013.

- Conrad, David E. "Tenant Farming and Sharecropping". Encyclopedia of Oklahoma History and Culture. Oklahoma Historical Society. Retrieved 16 September 2013.

5

Ecology: An Overview

The analysis of organisms and their environments is known as ecology. Ecologists seek to explain topics like life processes, adaptions and distribution of organisms. It is interdisciplinary and includes subjects like biology, geography and Earth science. This section is an overview of the subject matter incorporating all the major aspects of ecology, such as population ecology, restoration energy and ecological succession.

Ecology

Ecology ("house", "environment"; "study of") is the scien-tific analysis and study of interactions among organisms and their environment. It is an interdisciplinary field that includes biology, geography, and Earth science. Ecology includes the study of interactions organisms have with each other, other organisms, and with abiotic components of their environment. Topics of interest to ecologists include the diversity, distribution, amount (biomass), and number (population) of particular organisms, as well as cooperation and competition between organisms, both within and among ecosystems. Ecosystems are composed of dynamically interacting parts including organisms, the communities they make up, and the non-living components of their environment. Ecosystem processes, such as primary production, pedogenesis, nutrient cycling, and various niche construction activities, regulate the flux of energy and matter through an environment. These processes are sustained by organisms with specific life history traits, and the variety of organisms is called biodiversity. Biodiversity, which refers to the varieties of species, genes, and ecosystems, enhances certain ecosystem services.

Ecology is not synonymous with environment, environmentalism, natural history, or environmental science. It is closely related to evolutionary biology, genetics, and ethology. An important focus for ecologists is to improve the understanding of how biodiversity affects ecological function. Ecologists seek to explain:

- Life processes, interactions, and adaptations

- The movement of materials and energy through living communities

- The successional development of ecosystems

- The abundance and distribution of organisms and biodiversity in the context of the environment.

Ecology is a human science as well. There are many practical applications of ecology in conservation biology, wetland management, natural resource management (agroecology, agriculture, forestry, agroforestry, fisheries), city planning (urban ecology), community health, economics, basic and applied science, and human social interaction (human ecology). For example, the *Circles of Sustainability* approach treats ecology as more than the environment 'out there'. It is not treated as separate from humans. Organisms (including humans) and resources compose ecosystems which, in turn, maintain biophysical feedback mechanisms that moderate processes acting on living (biotic) and non-living (abiotic) components of the planet. Ecosystems sustain life-supporting functions and produce natural capital like biomass production (food, fuel, fiber, and medicine), the regulation of climate, global biogeochemical cycles, water filtration, soil formation, erosion control, flood protection, and many other natural features of scientific, historical, economic, or intrinsic value.

The word "ecology" ("Ökologie") was coined in 1866 by the German scientist Ernst Haeckel (1834–1919). Ecological thought is derivative of established currents in philosophy, particularly from ethics and politics. Ancient Greek philosophers such as Hippocrates and Aristotle laid the foundations of ecology in their studies on natural history. Modern ecology became a much more rigorous science in the late 19th century. Evolutionary concepts relating to adaptation and natural selection became the cornerstones of modern ecological theory.

Integrative Levels, Scope, and Scale of Organization

The scope of ecology contains a wide array of interacting levels of organization spanning micro-level (e.g., cells) to a planetary scale (e.g., biosphere) phenomena. Ecosystems, for example, contain abiotic resources and interacting life forms (i.e., individual organisms that aggregate into populations which aggregate into distinct ecological communities). Ecosystems are dynamic, they do not always follow a linear successional path, but they are always changing, sometimes rapidly and sometimes so slowly that it can take thousands of years for ecological processes to bring about certain successional stages of a forest. An ecosystem's area can vary greatly, from tiny to vast. A single tree is of little consequence to the classification of a forest ecosystem, but critically relevant to organisms living in and on it. Several generations of an aphid population can exist over the lifespan of a single leaf. Each of those aphids, in turn, support diverse bacterial communities. The nature of connections in ecological communities cannot be explained by knowing the details of each species in isolation, because the emergent pattern is neither revealed nor predicted until the ecosystem is studied as an integrated whole. Some ecological principles, however, do exhibit collective properties where the sum of the components explain the properties of the whole, such as birth rates of a population being equal to the sum of individual births over a designated time frame.

Hierarchical Ecology

System behaviors must first be arrayed into different levels of organization. Behaviors corresponding to higher levels occur at slow rates. Conversely, lower organizational levels exhibit rapid rates. For example, individual tree leaves respond rapidly to momentary changes in light intensity, CO_2 concentration, and the like. The growth of the tree responds more slowly and integrates these short-term changes.

O'Neill et al. (1986)

The scale of ecological dynamics can operate like a closed system, such as aphids migrating on a single tree, while at the same time remain open with regard to broader scale influences, such as atmosphere or climate. Hence, ecologists classify ecosystems hierarchically by analyzing data collected from finer scale units, such as vegetation associations, climate, and soil types, and integrate this information to identify emergent patterns of uniform organization and processes that operate on local to regional, landscape, and chronological scales.

To structure the study of ecology into a conceptually manageable framework, the biological world is organized into a nested hierarchy, ranging in scale from genes, to cells, to tissues, to organs, to organisms, to species, to populations, to communities, to ecosystems, to biomes, and up to the level of the biosphere. This framework forms a panarchy and exhibits non-linear behaviors; this means that "effect and cause are disproportionate, so that small changes to critical variables, such as the number of nitrogen fixers, can lead to disproportionate, perhaps irreversible, changes in the system properties."

Biodiversity

Biodiversity refers to the variety of life and its processes. It includes the variety of living organisms, the genetic differences among them, the communities and ecosystems in which they occur, and the ecological and evolutionary processes that keep them functioning, yet ever changing and adapting.

Noss & Carpenter (1994)

Biodiversity (an abbreviation of "biological diversity") describes the diversity of life from genes to ecosystems and spans every level of biological organization. The term has several interpretations, and there are many ways to index, measure, characterize, and represent its complex organization. Biodiversity includes species diversity, ecosystem diversity, and genetic diversity and scientists are interested in the way that this diversity affects the complex ecological processes operating at and among these respective levels. Biodiversity plays an important role in ecosystem services which by definition maintain and improve human quality of life. Preventing species extinctions is one way to preserve biodiversity and that goal rests on techniques that preserve genetic diversity, habitat and the ability for species to migrate. Conservation priorities and management techniques require different approaches and considerations to address the full ecological scope of biodiversity. Natural capital that supports populations is critical for maintaining ecosystem services and species migration (e.g., riverine fish runs and avian insect control) has been implicated as one mechanism by which those service losses are experienced. An understanding of biodiversity has practical applications for species and ecosystem-level conservation planners as they make management recommendations to consulting firms, governments, and industry.

Habitat

The habitat of a species describes the environment over which a species is known to occur and the type of community that is formed as a result. More specifically, "habitats can be defined as regions in environmental space that are composed of multiple dimensions, each representing a biotic or abiotic environmental variable; that is, any component or characteristic of the environment related directly (e.g. forage biomass and quality) or indirectly (e.g. elevation) to the use of a location by the animal." For example, a habitat might be an aquatic or terrestrial environment that can be further categorized as a montane or alpine ecosystem. Habitat shifts provide important evidence of competition in nature where one population changes relative to the habitats that most other individuals of the species occupy. For example, one population of a species of tropical lizards (*Tropidurus hispidus*) has a flattened body relative to the main populations that live in open savanna. The population that lives in an isolated rock outcrop hides in crevasses where its flattened body offers a selective advantage. Habitat shifts also occur in the developmental life history of amphibians, and in insects that transition from aquatic to terrestrial habitats. Biotope and habitat are sometimes used interchangeably, but the former applies to a community's environment, whereas the latter applies to a species' environment.

Additionally, some species are ecosystem engineers, altering the environment within a localized region. For instance, beavers manage water levels by building dams which improves their habitat in a landscape.

Biodiversity of a coral reef. Corals adapt to and modify their environment by forming calcium carbonate skeletons. This provides growing conditions for future generations and forms a habitat for many other species.

Niche

Termite mounds with varied heights of chimneys regulate gas exchange, temperature and other environmental parameters that are needed to sustain the internal physiology of the entire colony.

Definitions of the niche date back to 1917, but G. Evelyn Hutchinson made conceptual advances in 1957 by introducing a widely adopted definition: "the set of biotic and abiotic conditions in which a species is able to persist and maintain stable population sizes." The ecological niche is a central concept in the ecology of organisms and is sub-divided into the *fundamental* and the *realized* niche. The fundamental niche is the set of environmental conditions under which a species is able to persist. The realized niche is the set of environmental plus ecological conditions under which a species persists. The Hutchinsonian niche is defined more technically as a "Euclidean hyperspace whose *dimensions* are defined as environmental variables and whose *size* is a function of the number of values that the environmental values may assume for which an organism has *positive fitness.*"

Biogeographical patterns and range distributions are explained or predicted through knowledge of a species' traits and niche requirements. Species have functional traits that are uniquely adapted to the ecological niche. A trait is a measurable property, phenotype, or characteristic of an organism that may influence its survival. Genes play an important role in the interplay of development and environmental expression of traits. Resident species evolve traits that are fitted to the selection pressures of their local environment. This tends to afford them a competitive advantage and discourages similarly adapted species from having an overlapping geographic range. The competitive exclusion principle states that two species cannot coexist indefinitely by living off the same limiting resource; one will always out-compete the other. When similarly adapted species overlap geographically, closer inspection reveals subtle ecological differences in their habitat or dietary requirements. Some models and empirical studies, however, suggest that disturbances can stabilize the co-evolution and shared niche occupancy of similar species inhabiting species-rich communities. The habitat plus the niche is called the ecotope, which is defined as the full range of environmental and biological variables affecting an entire species.

Niche Construction

Organisms are subject to environmental pressures, but they also modify their habitats. The regulatory feedback between organisms and their environment can affect conditions from local (e.g., a beaver pond) to global scales, over time and even after death, such as decaying logs or silica skeleton deposits from marine organisms. The process and concept of ecosystem engineering is related to niche construction, but the former relates only to the physical modifications of the habitat whereas the latter also considers the evolutionary implications of physical changes to the environment and the feedback this causes on the process of natural selection. Ecosystem engineers are defined as: "organisms that directly or indirectly modulate the availability of resources to other species, by causing physical state changes in biotic or abiotic materials. In so doing they modify, maintain and create habitats."

The ecosystem engineering concept has stimulated a new appreciation for the influence that organisms have on the ecosystem and evolutionary process. The term "niche construction" is more often used in reference to the under-appreciated feedback mechanisms of natural selection imparting forces on the abiotic niche. An example of natural selection through ecosystem engineering occurs in the nests of social insects, including ants, bees, wasps, and termites. There is an emergent homeostasis or homeorhesis in the structure of the nest that regulates, maintains and defends the physiology of the entire colony. Termite mounds, for example, maintain a constant internal temperature through the design of air-conditioning chimneys. The structure of the nests themselves are subject to the forces of natural selection. Moreover, a nest can survive over successive generations, so that progeny inherit both genetic material and a legacy niche that was constructed before their time.

Biome

Biomes are larger units of organization that categorize regions of the Earth's ecosystems, mainly according to the structure and composition of vegetation. There are different methods to define the continental boundaries of biomes dominated by different functional types of vegetative communities that are limited in distribution by climate, precipitation, weather and other environmental variables. Biomes include tropical rainforest, temperate broadleaf and mixed forest, temperate deciduous forest, taiga, tundra, hot desert, and polar desert. Other researchers have recently categorized other biomes, such as the human and oceanic microbiomes. To a microbe, the human body is a habitat and a landscape. Microbiomes were discovered largely through advances in molecular genetics, which have revealed a hidden richness of microbial diversity on the planet. The oceanic microbiome plays a significant role in the ecological biogeochemistry of the planet's oceans.

Biosphere

The largest scale of ecological organization is the biosphere: the total sum of ecosystems on the planet. Ecological relationships regulate the flux of energy, nutrients, and climate all the way up to the planetary scale. For example, the dynamic history of the planetary atmosphere's CO_2 and O_2 composition has been affected by the biogenic flux of gases coming from respiration and photosynthesis, with levels fluctuating over time in relation to the ecology and evolution of plants and animals. Ecological theory has also been used to explain self-emergent regulatory phenomena at the planetary scale: for example, the Gaia hypothesis is an example of holism applied in ecological theory. The Gaia hypothesis states that there is an emergent feedback loop generated by the metabolism of living organisms that maintains the core temperature of the Earth and atmospheric conditions within a narrow self-regulating range of tolerance.

Individual Ecology

Understanding traits of individual organisms helps explain patterns and processes at other levels of organization including populations, communities, and ecosystems. Several areas of ecology of evolution that focus on such traits are life history theory, ecophysiology, metabolic theory of ecology, and Ethology. Examples of such traits include features of an organisms life cycle such as age to maturity, life span, or metabolic costs of reproduction. Other traits may be related to structure, such as the spines of a cactus or dorsal spines of a bluegill sunfish, or behaviors such as courtship displays or pair bonding. Other traits include emergent properties that are the result at least in part of interactions with the surrounding environment such as growth rate, resource uptake rate, winter, and deciduous vs. drought deciduous trees and shrubs.

One set of characteristics relate to body size and temperature. The metabolic theory of ecology provides a predictive qualitative set of relationships between an organism's body size and temperature and metabolic processes. In general, smaller, warmer organisms have higher metabolic rates and this results in a variety of predictions regarding individual somatic growth rates, reproduction and population growth rates, population size, and resource uptake rates.

The traits of organisms are subject to change through acclimation, development, and evolution. For this reason, individuals form a shared focus for ecology and for evolutionary ecology.

Population Ecology

Population ecology studies the dynamics of specie populations and how these populations interact with the wider environment. A population consists of individuals of the same species that live, interact, and migrate through the same niche and habitat.

A primary law of population ecology is the Malthusian growth model which states, "a population will grow (or decline) exponentially as long as the environment experienced by all individuals in the population remains constant." Simplified population models usually start with four variables: death, birth, immigration, and emigration.

An example of an introductory population model describes a closed population, such as on an island, where immigration and emigration does not take place. Hypotheses are evaluated with reference to a null hypothesis which states that random processes create the observed data. In these island models, the rate of population change is described by:

$$\frac{dN}{dT} = bN - dN = (b-d)N = rN,$$

where N is the total number of individuals in the population, b and d are the per capita rates of birth and death respectively, and r is the per capita rate of population change.

Using these modelling techniques, Malthus' population principle of growth was later transformed into a model known as the logistic equation:

$$\frac{dN}{dT} = aN\left(1 - \frac{N}{K}\right),$$

where N is the number of individuals measured as biomass density, a is the maximum per-capita rate of change, and K is the carrying capacity of the population. The formula states that the rate of change in population size (dN/dT) is equal to growth (aN) that is limited by carrying capacity ($1 - N/K$).

Population ecology builds upon these introductory models to further understand demographic processes in real study populations. Commonly used types of data include life history, fecundity, and survivorship, and these are analysed using mathematical techniques such as matrix algebra. The information is used for managing wildlife stocks and setting harvest quotas. In cases where basic models are insufficient, ecologists may adopt different kinds of statistical methods, such as the Akaike information criterion, or use models that can become mathematically complex as "several competing hypotheses are simultaneously confronted with the data."

Metapopulations and Migration

The concept of metapopulations was defined in 1969 as "a population of populations which go extinct locally and recolonize". Metapopulation ecology is another statistical approach that is often used in conservation research. Metapopulation models simplify the landscape into patches of varying levels of quality, and metapopulations are linked by the migratory behaviours of organisms. Animal migration is set apart from other kinds of movement; because, it involves the seasonal departure and return of individuals from a habitat. Migration is also a population-level phenomenon, as with the migration routes followed by plants as they occupied northern post-glacial environments. Plant ecologists use pollen records that accumulate and stratify in wetlands to reconstruct the timing of plant migration and dispersal relative to historic and contemporary climates. These migration routes involved an expansion of the range as plant populations expanded from one area to another. There is a larger taxonomy of movement, such as commuting, foraging, territorial behaviour, stasis, and ranging. Dispersal is usually distinguished from migration; because, it involves the one way permanent movement of individuals from their birth population into another population.

In metapopulation terminology, migrating individuals are classed as emigrants (when they leave a region) or immigrants (when they enter a region), and sites are classed either as sources or sinks. A site is a generic term that refers to places where ecologists sample populations, such as ponds or defined sampling areas in a forest. Source patches are productive sites that generate a seasonal supply of juveniles that migrate to other patch locations. Sink patches are unproductive sites that only receive migrants; the population at the site will disappear unless rescued by an adjacent source patch or environmental conditions become more favourable. Metapopulation models examine patch dynamics over time to answer potential questions about spatial and demographic ecology. The ecology of metapopulations is a dynamic process of extinction and colonization. Small patches of lower quality (i.e., sinks) are maintained or rescued by a seasonal influx of new immigrants. A dynamic metapopulation structure evolves from year to year, where some patches are sinks in dry years and are sources when conditions are more favourable. Ecologists use a mixture of computer models and field studies to explain metapopulation structure.

Community Ecology

Interspecific interactions such as predation are a key aspect of community ecology.

Community ecology examines how interactions among species and their environment affect the abundance, distribution and diversity of species within communities.

Johnson & Stinchcomb (2007)

Community ecology is the study of the interactions among a collections of species that inhabit the same geographic area. Community ecologists study the determinants of patterns and processes for two or more interacting species. Research in community ecology might measure species diversity in grasslands in relation to soil fertility. It might also include the analysis of predator-prey dynamics, competition among similar plant species, or mutualistic interactions between crabs and corals.

Ecosystem ecology

These ecosystems, as we may call them, are of the most various kinds and sizes. They form one category of the multitudinous physical systems of the universe, which range from the universe as a whole down to the atom.

Tansley (1935)

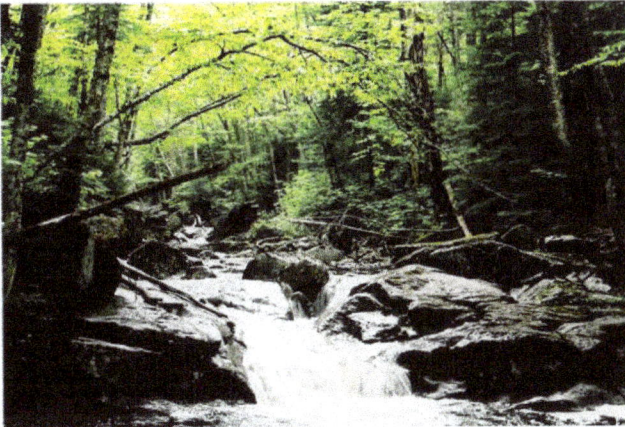

A riparian forest in the White Mountains, New Hampshire (USA), an example of ecosystem ecology

Ecosystems may be habitats within biomes that form an integrated whole and a dynamically responsive system having both physical and biological complexes. Ecosystem ecology is the science of determining the fluxes of materials (e.g. carbon, phosphorus) between different pools (e.g., tree biomass, soil organic material). Ecosystem ecologist attempt to determine the underlying causes of these fluxes. Research in ecosystem ecology might measure primary production (g C/m^2) in a wetland in relation to decomposition and consumption rates (g C/m^2/y). This requires an understanding of the community connections between plants (i.e., primary producers) and the decomposers (e.g., fungi and bacteria),

The underlying concept of ecosystem can be traced back to 1864 in the published work of George Perkins Marsh ("Man and Nature"). Within an ecosystem, organisms are linked to the physical and biological components of their environment to which they are adapted. Ecosystems are complex adaptive systems where the interaction of life processes form self-organizing patterns across different scales of time and space. Ecosystems are broadly categorized as terrestrial, freshwater, atmospheric, or marine. Differences stem from the nature of the unique physical environments that shapes the biodiversity within each. A more recent addition to ecosystem ecology are technoecosystems, which are affected by or primarily the result of human activity.

Food Webs

A food web is the archetypal ecological network. Plants capture solar energy and use it to synthesize simple sugars during photosynthesis. As plants grow, they accumulate nutrients and are eaten by grazing herbivores, and the energy is transferred through a chain of organisms by consumption. The simplified linear feeding pathways that move from a basal trophic species to a top consumer is called the food chain. The larger inter-locking pattern of food chains in an ecological community creates a complex food web. Food webs are a type of concept map or a heuristic device that is used to illustrate and study pathways of energy and material flows.

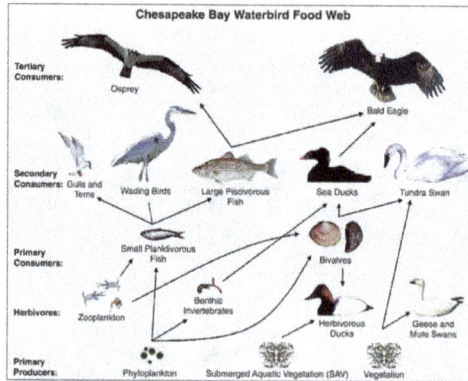

Generalized food web of waterbirds from Chesapeake Bay

Food webs are often limited relative to the real world. Complete empirical measure-ments are generally restricted to a specific habitat, such as a cave or a pond, and prin-ciples gleaned from food web microcosm studies are extrapolated to larger systems. Feeding relations require extensive investigations into the gut contents of organisms, which can be difficult to decipher, or stable isotopes can be used to trace the flow of nu-trient diets and energy through a food web. Despite these limitations, food webs remain a valuable tool in understanding community ecosystems.

Food webs exhibit principles of ecological emergence through the nature of trophic rela-tionships: some species have many weak feeding links (e.g., omnivores) while some are more specialized with fewer stronger feeding links (e.g., primary predators). Theoretical and empirical studies identify non-random emergent patterns of few strong and many weak linkages that explain how ecological communities remain stable over time. Food webs are composed of subgroups where members in a community are linked by strong interactions, and the weak interactions occur between these subgroups. This increases food web stability. Step by step lines or relations are drawn until a web of life is illustrated.

Trophic Levels

A trophic level (from Greek *troph*, τροφή, trophē, meaning "food" or "feeding") is "a group of organisms acquiring a considerable majority of its energy from the adjacent level nearer the abiotic source." Links in food webs primarily connect feeding relations or trophism among species. Biodiversity within ecosystems can be organized into trophic pyramids, in which the vertical dimension represents feeding relations that become further removed from the base of the food chain up toward top predators, and the horizontal dimension represents the abundance or biomass at each level. When the relative abundance or biomass of each species is sorted into its respective trophic level, they naturally sort into a 'pyramid of numbers'.

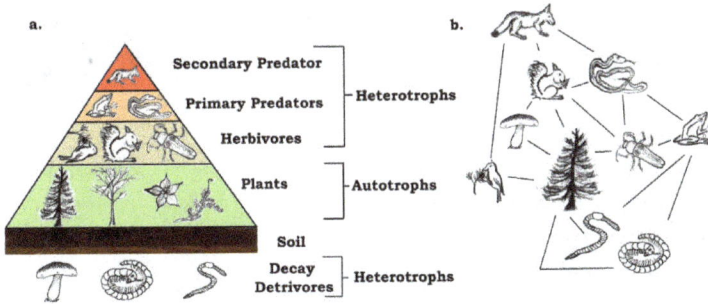

A trophic pyramid (a) and a food-web (b) illustrating ecological relationships among creatures that are typical of a northern boreal terrestrial ecosystem. The trophic pyramid roughly represents the biomass (usually measured as total dry-weight) at each level. Plants generally have the greatest biomass. Names of trophic categories are shown to the right of the pyramid. Some ecosystems, such as many wetlands, do not organize as a strict pyramid, because aquatic plants are not as productive as long-lived terrestrial plants such as trees. Ecological trophic pyramids are typically one of three kinds: 1) pyramid of numbers, 2) pyramid of biomass, or 3) pyramid of energy.

Species are broadly categorized as autotrophs (or primary producers), heterotrophs (or consumers), and Detritivores (or decomposers). Autotrophs are organisms that produce their own food (production is greater than respiration) by photosynthesis or chemosynthesis. Heterotrophs are organisms that must feed on others for nourishment and energy (respiration exceeds production). Heterotrophs can be further sub-divided into different functional groups, including primary consumers (strict herbivores), secondary consumers (carnivorous predators that feed exclusively on herbivores), and tertiary consumers (predators that feed on a mix of herbivores and predators). Omnivores do not fit neatly into a functional category because they eat both plant and animal tissues. It has been suggested that omnivores have a greater functional influence as predators, because compared to herbivores, they are relatively inefficient at grazing.

Trophic levels are part of the holistic or complex systems view of ecosystems. Each trophic level contains unrelated species that are grouped together because they share common ecological functions, giving a macroscopic view of the system. While the notion of trophic levels provides insight into energy flow and top-down control within food webs, it is troubled by the prevalence of omnivory in real ecosystems. This has led some ecologists to "reiterate that the notion that species clearly aggregate into discrete, homogeneous trophic levels is fiction." Nonetheless, recent studies have shown that real trophic levels do exist, but "above the herbivore trophic level, food webs are better characterized as a tangled web of omnivores."

Keystone Species

Sea otters, an example of a keystone species

A keystone species is a species that is connected to a disproportionately large number of other species in the food-web. Keystone species have lower levels of biomass in the trophic pyramid relative to the importance of their role. The many connections that a keystone species holds means that it maintains the organization and structure of entire communities. The loss of a keystone species results in a range of dramatic cascading effects that alters trophic dynamics, other food web connections, and can cause the extinction of other species.

Sea otters (*Enhydra lutris*) are commonly cited as an example of a keystone species; because, they limit the density of sea urchins that feed on kelp. If sea otters are removed from the system, the urchins graze until the kelp beds disappear, and this has a dramatic effect on community structure. Hunting of sea otters, for example, is thought to have led indirectly to the extinction of the Steller's sea cow (*Hydrodamalis gigas*). While the keystone species concept has been used extensively as a conservation tool, it has been criticized for being poorly defined from an operational stance. It is difficult to experimentally determine what species may hold a keystone role in each ecosystem. Furthermore, food web theory suggests that keystone species may not be common, so it is unclear how generally the keystone species model can be applied.

Ecological Complexity

Complexity is understood as a large computational effort needed to piece together numerous interacting parts exceeding the iterative memory capacity of the human mind. Global patterns of biological diversity are complex. This biocomplexity stems from the interplay among ecological processes that operate and influence patterns at different scales that grade into each other, such as transitional areas or ecotones spanning landscapes. Complexity stems from the interplay among levels of biological organization as energy, and matter is integrated into larger units that superimpose onto the smaller parts. "What were wholes on one level become parts on a higher one." Small scale patterns do not necessarily explain large scale phenomena, otherwise captured in the expression (coined by Aristotle) 'the sum is greater than the parts'.

"Complexity in ecology is of at least six distinct types: spatial, temporal, structural, process, behavioral, and geometric." From these principles, ecologists have identified emergent and self-organizing phenomena that operate at different environmental scales of influence, ranging from molecular to planetary, and these require different explanations at each integrative level. Ecological complexity relates to the dynamic resilience of ecosystems that transition to multiple shifting steady-states directed by random fluctuations of history. Long-term ecological studies provide important track records to better understand the complexity and resilience of ecosystems over longer temporal and broader spatial scales. These studies are managed by the International Long Term Ecological Network (LTER). The longest experiment in existence is the Park Grass Experiment, which was initiated in 1856. Another example is the Hubbard Brook study, which has been in operation since 1960.

Holism

Holism remains a critical part of the theoretical foundation in contemporary ecological studies. Holism addresses the biological organization of life that self-organizes into layers of emergent whole systems that function according to non-reducible properties. This means that higher order patterns of a whole functional system, such as an ecosystem, cannot be predicted or understood by a simple summation of the parts. "New properties emerge because the components interact, not because the basic nature of the components is changed."

Ecological studies are necessarily holistic as opposed to reductionistic. Holism has three scientific meanings or uses that identify with ecology: 1) the mechanistic complexity of ecosystems, 2) the practical description of patterns in quantitative reductionist terms where correlations may be identified but nothing is understood about the causal relations without reference to the whole system, which leads to 3) a metaphysical hierarchy whereby the causal relations of larger systems are understood without reference to the smaller parts. Scientific holism differs from mysticism that has appropriated the same term. An example of metaphysical holism is identified in the trend of increased exterior thickness in shells of different species. The reason for a thickness increase can be understood through reference to principles of natural selection via predation without need to reference or understand the biomolecular properties of the exterior shells.

Relation to Evolution

Ecology and evolution are considered sister disciplines of the life sciences. Natural selection, life history, development, adaptation, populations, and inheritance are examples of concepts that thread equally into ecological and evolutionary theory. Morphological, behavioural, and genetic traits, for example, can be mapped onto evolutionary trees to study the historical development of a species in relation to their functions and roles in different ecological circumstances. In this framework, the analytical tools of ecologists and evolutionists overlap as they organize, classify, and investigate life through common systematic principals, such as phylogenetics or the Linnaean system of taxonomy. The two disciplines often appear together, such as in the title of the journal *Trends in Ecology and Evolution*. There is no sharp boundary separating ecology from evolution, and they differ more in their areas of applied focus. Both disciplines discover and explain emergent and unique properties and processes operating across different spatial or temporal scales of organization. While the boundary between ecology and evolution is not always clear, ecologists study the abiotic and biotic factors that influence evolutionary processes, and evolution can be rapid, occurring on ecological timescales as short as one generation.

Behavioural Ecology

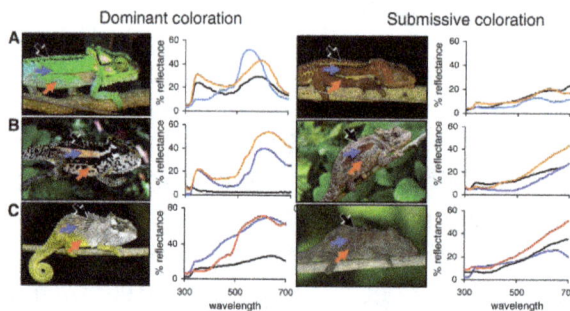

Social display and colour variation in differently adapted species of chameleons (*Bradypodion* spp.). Chameleons change their skin colour to match their background as a behavioural defence mechanism and also use colour to communicate with other members of their species, such as dominant (left) versus submissive (right) patterns shown in the three species (A-C) above.

All organisms can exhibit behaviours. Even plants express complex behaviour, including memory and communication. Behavioural ecology is the study of an organism's behaviour in its environment and its ecological and evolutionary implications. Ethology is the study of observable movement or behaviour in animals. This could include investigations of motile sperm of plants, mobile phytoplankton, zooplankton swimming toward the female egg, the cultivation of fungi by weevils, the mating dance of a salamander, or social gatherings of amoeba.

Adaptation is the central unifying concept in behavioural ecology. Behaviours can be recorded as traits and inherited in much the same way that eye and hair colour can. Behaviours can evolve by means of natural selection as adaptive traits conferring functional utilities that increases reproductive fitness.

Predator-prey interactions are an introductory concept into food-web studies as well as behavioural ecology. Prey species can exhibit different kinds of behavioural adaptations to predators, such as avoid, flee, or defend. Many prey species are faced with multiple predators that differ in the degree of danger posed. To be adapted to their environment and face predatory threats, organisms must balance their energy budgets as they invest in different aspects of their life history, such as growth, feeding, mating, socializing, or modifying their habitat. Hypotheses posited in behavioural ecology are generally based on adaptive principles of conservation, optimization, or efficiency. For example, "[t]he threat-sensitive predator avoidance hypothesis predicts that prey should assess the degree of threat posed by different predators and match their behaviour according to current levels of risk" or "[t]he optimal flight initiation distance occurs where expected postencounter fitness is maximized, which depends on the prey's initial fitness, benefits obtainable by not fleeing, energetic escape costs, and expected fitness loss due to predation risk."

Symbiosis: Leafhoppers (*Eurymela fenestrata*) are protected by ants (*Iridomyrmex purpureus*) in a symbiotic relationship. The ants protect the leafhoppers from predators and in return the leafhoppers feeding on plants exude honeydew from their anus that provides energy and nutrients to tending ants.

Elaborate sexual displays and posturing are encountered in the behavioural ecology of animals. The birds-of-paradise, for example, sing and display elaborate ornaments during courtship. These displays serve a dual purpose of signalling healthy or well-adapted individuals and desirable genes. The displays are driven by sexual selection as an advertisement of quality of traits among suitors.

Cognitive Ecology

Cognitive ecology integrates theory and observations from evolutionary ecology and neurobiology, primarily cognitive science, in order to understand the effect that animal interaction with their habitat has on their cognitive systems and how those systems restrict behavior within an ecological and evolutionary framework. "Until recently, however, cognitive scientists have not paid sufficient attention to the fundamental fact that cognitive traits evolved under particular natural settings. With consideration of the selection pressure on cognition, cognitive ecology can contribute intellectual coherence to the multidisciplinary study of cognition." As a study involving the 'coupling' or interactions between organism and environment, cognitive ecology is closely related to enactivism, a field based upon the view that "...we must see the organism and environment as bound together in reciprocal specification and selection...".

Social Ecology

Social ecological behaviours are notable in the social insects, slime moulds, social spiders, human society, and naked mole-rats where eusocialism has evolved. Social behaviours include reciprocally beneficial behaviours among kin and nest mates and evolve from kin and group selection. Kin selection explains altruism through genetic relationships, whereby an altruistic behaviour leading to death is rewarded by the survival of genetic copies distributed among surviving relatives. The social insects, including ants, bees, and wasps are most famously studied for this type of relationship because the male drones are clones that share the same genetic make-up as every other male in the colony. In contrast, group selectionists find examples of altruism among non-genetic relatives and explain this through selection acting on the group; whereby, it becomes selectively advantageous for groups if their members express altruistic behaviours to one another. Groups with predominantly altruistic members beat groups with predominantly selfish members.

Coevolution

Bumblebees and the flowers they pollinate have coevolved so that both have become dependent on each other for survival.

Ecological interactions can be classified broadly into a host and an associate relationship. A host is any entity that harbours another that is called the associate. Relationships within a species that are mutually or reciprocally beneficial are called mutualisms. Examples of mutualism include fungus-growing ants employing agricultural symbiosis, bacteria living in the guts of insects and other organisms, the fig wasp and yucca moth pollination complex, lichens with fungi and photosynthetic algae, and corals with photosynthetic algae. If there is a physical connection between host and associate, the relationship is called symbiosis. Approximately 60% of all plants, for example, have a symbiotic relationship with arbuscular mycorrhizal fungi living in their roots forming an exchange network of carbohydrates for mineral nutrients.

Indirect mutualisms occur where the organisms live apart. For example, trees living in the equatorial regions of the planet supply oxygen into the atmosphere that sustains species living in distant polar regions of the planet. This relationship is called commensalism; because, many others receive the benefits of clean air at no cost or harm to trees supplying the oxygen. If the associate benefits while the host suffers, the relationship is called parasitism. Although parasites impose a cost to their host (e.g., via damage to their reproductive organs or propagules, denying the services of a beneficial partner), their net effect on host fitness is not necessarily negative and, thus, becomes difficult to forecast. Co-evolution is also driven by competition among species or among members of the same species under the banner of reciprocal antagonism, such as grasses competing for growth space. The Red Queen Hypothesis, for example, posits that parasites track down and specialize on the locally common genetic defense systems of its host that drives the evolution of sexual reproduction to diversify the genetic constituency of populations responding to the antagonistic pressure.

Parasitism: A harvestman arachnid being parasitized by mites. The harvestman is being consumed, while the mites benefit from traveling on and feeding off of their host.

Biogeography

Biogeography (an amalgamation of *biology* and *geography*) is the comparative study of the geographic distribution of organisms and the corresponding evolution of their traits in space and time. The *Journal of Biogeography* was established in 1974. Biogeography and ecology share many of their disciplinary roots. For example, the theory of island biogeography, published by the mathematician Robert MacArthur and ecologist Edward O. Wilson in 1967 is considered one of the fundamentals of ecological theory.

Biogeography has a long history in the natural sciences concerning the spatial distribution of plants and animals. Ecology and evolution provide the explanatory context for biogeographical studies. Biogeographical patterns result from ecological processes that influence range distributions, such as migration and dispersal. and from historical processes that split populations or species into different areas. The biogeographic processes that result in the natural splitting of species explains much of the modern distribution of the Earth's biota. The splitting of lineages in a species is called vicariance biogeography and it is a sub-discipline of biogeography. There are also practical applications in the field of biogeography concerning ecological systems and processes. For example, the range and distribution of biodiversity and invasive species responding to climate change is a serious concern and active area of research in the context of global warming.

R/K-Selection Theory

A population ecology concept is r/K selection theory, one of the first predictive models in ecology used to explain life-history evolution. The premise behind the r/K selection model is that natural selection pressures change according to population density. For example, when an island is first colonized, density of individuals is low. The initial increase in population size is not limited by competition, leaving an abundance of available resources for rapid population growth. These early phases of population growth experience *density-independent* forces of natural selection, which is called r-selection. As the population becomes more crowded, it approaches the island's carrying capacity, thus forcing individuals to compete more heavily for fewer available resources. Under crowded conditions, the population experiences density-dependent forces of natural selection, called K-selection.

In the r/K-selection model, the first variable r is the intrinsic rate of natural increase in population size and the second variable K is the carrying capacity of a population. Different species evolve different life-history strategies spanning a continuum between these two selective forces. An r-selected species is one that has high birth rates, low levels of parental investment, and high rates of mortality before individuals reach maturity. Evolution favours high rates of fecundity in r-selected species. Many kinds of insects and invasive species exhibit r-selected characteristics. In contrast, a K-selected species has low rates of fecundity, high levels of parental investment in the young, and low rates of mortality as individuals mature. Humans and elephants are examples of species exhibiting K-selected characteristics, including longevity and efficiency in the conversion of more resources into fewer offspring.

Molecular Ecology

The important relationship between ecology and genetic inheritance predates modern techniques for molecular analysis. Molecular ecological research became more feasible with the development of rapid and accessible genetic technologies, such as the polymerase chain reaction (PCR). The rise of molecular technologies and influx of research questions into this new ecological field resulted in the publication *Molecular Ecology* in 1992. Molecular ecology uses various analytical techniques to study genes in an evolutionary and ecological context. In 1994, John Avise also played a leading role in this area of science with the publication of his book, *Molecular Markers, Natural History and Evolution*. Newer technologies opened a wave of genetic analysis into organisms once difficult to study from an ecological or evolutionary standpoint, such as bacteria, fungi, and nematodes. Molecular ecology engendered a new research paradigm for investigating ecological questions considered otherwise intractable. Molecular investigations revealed previously obscured details in the tiny intricacies of nature and improved resolution into probing questions about behavioural and biogeographical ecology. For example, molecular ecology revealed promiscuous sexual behaviour and multiple male partners in tree swallows previously thought to be socially monogamous. In a biogeographical context, the marriage between genetics, ecology, and evolution resulted in a new sub-discipline called phylogeography.

Human Ecology

The history of life on Earth has been a history of interaction between living things and their surroundings. To a large extent, the physical form and the habits of the earth's vegetation and its animal life have been molded by the environment. Considering the whole span of earthly time, the opposite effect, in which life actually modifies its surroundings, has been relatively slight. Only within the moment of time represented by the present century has one species man acquired significant power to alter the nature of his world.

Rachel Carson, "Silent Spring"

Ecology is as much a biological science as it is a human science. Human ecology is an interdisciplinary investigation into the ecology of our species. "Human ecology may be defined: (1) from a bio-ecological standpoint as the study of man as the ecological dominant in plant and animal communities and systems; (2) from a bio-ecological standpoint as simply another animal affecting and being affected by his physical environment; and (3) as a human being, somehow different from animal life in general, interacting with physical and modified environments in a distinctive and creative way. A truly interdisciplinary human ecology will most likely address itself to all three." The term was formally introduced in 1921, but many sociologists, geographers, psychologists, and other disciplines were interested in human relations to natural systems centuries prior, especially in the late 19th century.

The ecological complexities human beings are facing through the technological transformation of the planetary biome has brought on the Anthropocene. The unique set of circumstances has generated the need for a new unifying science called coupled human and natural systems that builds upon, but moves beyond the field of human ecology. Ecosystems tie into human societies through the critical and all encompassing life-supporting functions they sustain. In recognition of these functions and the incapability of traditional economic valuation methods to see the value in ecosystems, there has been a surge of interest in social-natural capital, which provides the means to put a value on the stock and use of information and materials stemming from ecosystem goods and services. Ecosystems produce, regulate, maintain, and supply services of critical necessity and beneficial to human health (cognitive and physiological), economies, and they even provide an information or reference function as a living library giving opportunities for science and cognitive development in children engaged in the complexity of the natural world. Ecosystems relate importantly to human ecology as they are the ultimate base foundation of global economics as every commodity, and the capacity for exchange ultimately stems from the ecosystems on Earth.

Restoration and Management

Ecosystem management is not just about science nor is it simply an extension of traditional resource management; it offers a fundamental reframing of how humans may work with nature.

Grumbine (1994)

Ecology is an employed science of restoration, repairing disturbed sites through human intervention, in natural resource management, and in environmental impact assessments. Edward O. Wilson predicted in 1992 that the 21st century "will be the era of restoration in ecology". Ecological science has boomed in the industrial investment of restoring ecosystems and their processes in abandoned sites after disturbance. Natural resource managers, in forestry, for example, employ ecologists to develop, adapt, and implement ecosystem based methods into the planning, operation, and restoration phases of land-use. Ecological science is used in the methods of sustainable harvesting, disease, and fire outbreak management, in fisheries stock management, for integrating land-use with protected areas and communities, and conservation in complex geo-political landscapes.

Relation to the Environment

The environment of ecosystems includes both physical parameters and biotic attributes. It is dynamically interlinked, and contains resources for organisms at any time throughout their life cycle. Like "ecology", the term "environment" has different conceptual meanings and overlaps with the concept of "nature". Environment "includes the physical world, the social world of human relations and the built world of human creation." The physical environment is external to the level of biological organization under investigation, including abiotic factors such as temperature, radiation, light, chemistry, climate and geology. The biotic environment includes genes, cells, organisms, members of the same species (conspecifics) and other species that share a habitat.

The distinction between external and internal environments, however, is an abstraction parsing life and environment into units or facts that are inseparable in reality. There is an interpenetration of cause and effect between the environment and life. The laws of thermodynamics, for example, apply to ecology by means of its physical state. With an understanding of metabolic and thermodynamic principles, a complete accounting of energy and material flow can be traced through an ecosystem. In this way, the environmental and ecological relations are studied through reference to conceptually manageable and isolated material parts. After the effective environmental components are understood through reference to their causes; however, they conceptually link back together as an integrated whole, or *holocoenotic* system as it was once called. This is known as the dialectical approach to ecology. The dialectical approach examines the parts, but integrates the organism and the environment into a dynamic whole (or umwelt). Change in one ecological or environmental factor can concurrently affect the dynamic state of an entire ecosystem.

Disturbance and Resilience

Ecosystems are regularly confronted with natural environmental variations and disturbances over time and geographic space. A disturbance is any process that removes biomass from a community, such as a fire, flood, drought, or predation. Disturbances occur over vastly different ranges in terms of magnitudes as well as distances and time periods, and are both the cause and product of natural fluctuations in death rates, species assemblages, and biomass densities within an ecological community. These disturbances create places of renewal where new directions emerge from the patchwork of natural experimentation and opportunity. Ecological resilience is a cornerstone theory in ecosystem management. Biodiversity fuels the resilience of ecosystems acting as a kind of regenerative insurance.

Metabolism and the Early Atmosphere

Metabolism – the rate at which energy and material resources are taken up from the environment, transformed within an organism, and allocated to maintenance, growth and reproduction – is a fundamental physiological trait.

Ernest et al.

The Earth was formed approximately 4.5 billion years ago. As it cooled and a crust and oceans formed, its atmosphere transformed from being dominated by hydrogen to one composed mostly of methane and ammonia. Over the next billion years, the metabolic activity of life transformed the atmosphere into a mixture of carbon dioxide, nitrogen, and water vapor. These gases changed the way that light from the sun hit the Earth's surface and greenhouse effects trapped heat. There were untapped sources of free energy within the mixture of reducing and oxidizing gasses that set the stage for primitive ecosystems to evolve and, in turn, the atmosphere also evolved.

The leaf is the primary site of photosynthesis in most plants.

Throughout history, the Earth's atmosphere and biogeochemical cycles have been in a dynamic equilibrium with planetary ecosystems. The history is characterized by periods of significant transformation followed by millions of years of stability. The evolution of the earliest organisms, likely anaerobic methanogen microbes, started the process by converting atmospheric hydrogen into methane ($4H_2 + CO_2 \rightarrow CH_4 + 2H_2O$). Anoxygenic photosynthesis reduced hydrogen concentrations and increased atmospheric methane, by converting hydrogen sulfide into water or other sulfur compounds (for example, $2H_2S + CO_2 + h\upsilon \rightarrow CH_2O + H_2O + 2S$). Early forms of fermentation also increased levels of atmospheric methane. The transition to an oxygen-dominant atmosphere (the *Great Oxidation*) did not begin until approximately 2.4–2.3 billion years ago, but photosynthetic processes started 0.3 to 1 billion years prior.

Radiation: Heat, Temperature and Light

The biology of life operates within a certain range of temperatures. Heat is a form of energy that regulates temperature. Heat affects growth rates, activity, behaviour, and primary production. Temperature is largely dependent on the incidence of solar radiation. The latitudinal and longitudinal spatial variation of temperature greatly affects climates and consequently the distribution of biodiversity and levels of primary production in different ecosystems or biomes across the planet. Heat and temperature relate importantly to metabolic activity. Poikilotherms, for example, have a body temperature that is largely regulated and dependent on the temperature of the external environment. In contrast, homeotherms regulate their internal body temperature by expending metabolic energy.

There is a relationship between light, primary production, and ecological energy budgets. Sunlight is the primary input of energy into the planet's ecosystems. Light is composed of electromagnetic energy of different wavelengths. Radiant energy from the sun generates heat, provides photons of light measured as active energy in the chemical reactions of life, and also acts as a catalyst for genetic mutation. Plants, algae, and some bacteria absorb light and assimilate the energy through photosynthesis. Organisms capable of assimilating energy by photosynthesis or through inorganic fixation of H_2S are autotrophs. Autotrophs — responsible for primary production — assimilate light energy which becomes metabolically stored as potential energy in the form of biochemical enthalpic bonds.

Physical Environments

Water

Wetland conditions such as shallow water, high plant productivity, and anaerobic substrates provide a suitable environment for important physical, biological, and chemical processes. Because of these processes, wetlands play a vital role in global nutrient and element cycles.

Cronk & Fennessy (2001)

Diffusion of carbon dioxide and oxygen is approximately 10,000 times slower in water than in air. When soils are flooded, they quickly lose oxygen, becoming hypoxic (an environment with O_2 concentration below 2 mg/liter) and eventually completely anoxic where anaerobic bacteria thrive among the roots. Water also influences the intensity and spectral composition of light as it reflects off the water surface and submerged particles. Aquatic plants exhibit a wide variety of morphological and physiological adaptations that allow them to survive, compete, and diversify in these environments. For example, their roots and stems contain large air spaces (aerenchyma) that regulate the efficient transportation of gases (for example, CO_2 and O_2) used in respiration and photosynthesis. Salt water plants (halophytes) have additional specialized adaptations, such as the development of special organs for shedding salt and osmoregulating their internal salt ($NaCl$) concentrations, to live in estuarine, brackish, or oceanic environments. Anaerobic soil microorganisms in aquatic environments use nitrate, manganese ions, ferric ions, sulfate, carbon dioxide, and some organic compounds; other microorganisms are facultative anaerobes and use oxygen during respiration when the soil becomes drier. The activity of soil microorganisms and the chemistry of the water reduces the oxidation-reduction potentials of the water. Carbon dioxide, for example, is reduced to methane (CH_4) by methanogenic bacteria. The physiology of fish is also specially adapted to compensate for environmental salt levels through osmoregulation. Their gills form electrochemical gradients that mediate salt excretion in salt water and uptake in fresh water.

Gravity

The shape and energy of the land is significantly affected by gravitational forces. On a large scale, the distribution of gravitational forces on the earth is uneven and influences the shape and movement of tectonic plates as well as influencing geomorphic processes such as orogeny and erosion. These forces govern many of the geophysical properties and distributions of ecological biomes across the Earth. On the organismal scale, gravitational forces provide directional cues for plant and fungal growth (gravitropism), orientation cues for animal migrations, and influence the biomechanics and size of an-

imals. Ecological traits, such as allocation of biomass in trees during growth are subject to mechanical failure as gravitational forces influence the position and structure of
branches and leaves. The cardiovascular systems of animals are functionally adapted
to overcome pressure and gravitational forces that change according to the features of
organisms (e.g., height, size, shape), their behaviour (e.g., diving, running, flying), and
the habitat occupied (e.g., water, hot deserts, cold tundra).

Pressure

Climatic and osmotic pressure places physiological constraints on organisms, especially those that fly and respire at high altitudes, or dive to deep ocean depths. These
constraints influence vertical limits of ecosystems in the biosphere, as organisms are
physiologically sensitive and adapted to atmospheric and osmotic water pressure differences. For example, oxygen levels decrease with decreasing pressure and are a limiting factor for life at higher altitudes. Water transportation by plants is another important ecophysiological process affected by osmotic pressure gradients. Water pressure in
the depths of oceans requires that organisms adapt to these conditions. For example,
diving animals such as whales, dolphins, and seals are specially adapted to deal with
changes in sound due to water pressure differences. Differences between hagfish species provide another example of adaptation to deep-sea pressure through specialized
protein adaptations.

Wind and Turbulence

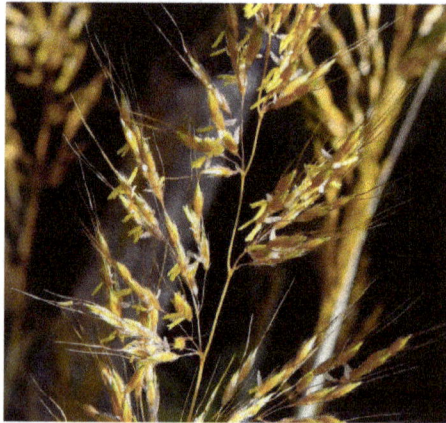

The architecture of the inflorescence in grasses is subject to the physical pressures of wind and shaped by
the forces of natural selection facilitating wind-pollination (anemophily).

Turbulent forces in air and water affect the environment and ecosystem distribution,
form and dynamics. On a planetary scale, ecosystems are affected by circulation patterns in the global trade winds. Wind power and the turbulent forces it creates can
influence heat, nutrient, and biochemical profiles of ecosystems. For example, wind
running over the surface of a lake creates turbulence, mixing the water column and
influencing the environmental profile to create thermally layered zones, affecting how

fish, algae, and other parts of the aquatic ecosystem are structured. Wind speed and turbulence also influence evapotranspiration rates and energy budgets in plants and animals. Wind speed, temperature and moisture content can vary as winds travel across different land features and elevations. For example, the westerlies come into contact with the coastal and interior mountains of western North America to produce a rain shadow on the leeward side of the mountain. The air expands and moisture condenses as the winds increase in elevation; this is called orographic lift and can cause precipitation. This environmental process produces spatial divisions in biodiversity, as species adapted to wetter conditions are range-restricted to the coastal mountain valleys and unable to migrate across the xeric ecosystems (e.g., of the Columbia Basin in western North America) to intermix with sister lineages that are segregated to the interior mountain systems.

Fire

Forest fires modify the land by leaving behind an environmental mosaic that diversifies the landscape into different seral stages and habitats of varied quality (left). Some species are adapted to forest fires, such as pine trees that open their cones only after fire exposure (right).

Plants convert carbon dioxide into biomass and emit oxygen into the atmosphere. By approximately 350 million years ago (the end of the Devonian period), photosynthesis had brought the concentration of atmospheric oxygen above 17%, which allowed combustion to occur. Fire releases CO_2 and converts fuel into ash and tar. Fire is a significant ecological parameter that raises many issues pertaining to its control and suppression. While the issue of fire in relation to ecology and plants has been recognized for a long time, Charles Cooper brought attention to the issue of forest fires in relation to the ecology of forest fire suppression and management in the 1960s.

Native North Americans were among the first to influence fire regimes by controlling their spread near their homes or by lighting fires to stimulate the production of herbaceous foods and basketry materials. Fire creates a heterogeneous ecosystem age and canopy structure, and the altered soil nutrient supply and cleared canopy structure opens new ecological niches for seedling establishment. Most ecosystems are adapted to natural fire cycles. Plants, for example, are equipped with a variety of adaptations to deal with forest fires. Some species (e.g., *Pinus halepensis*) cannot germinate until after their seeds have lived through a fire or been exposed to certain compounds from smoke. Environmentally triggered germination of seeds is called serotiny. Fire plays a major role in the persistence and resilience of ecosystems.

Soils

Soil is the living top layer of mineral and organic dirt that covers the surface of the planet. It is the chief organizing centre of most ecosystem functions, and it is of critical importance in agricultural science and ecology. The decomposition of dead organic matter (for example, leaves on the forest floor), results in soils containing minerals and nutrients that feed into plant production. The whole of the planet's soil ecosystems is called the pedosphere where a large biomass of the Earth's biodiversity organizes into trophic levels. Invertebrates that feed and shred larger leaves, for example, create smaller bits for smaller organisms in the feeding chain. Collectively, these organisms are the detritivores that regulate soil formation. Tree roots, fungi, bacteria, worms, ants, beetles, centipedes, spiders, mammals, birds, reptiles, amphibians, and other less familiar creatures all work to create the trophic web of life in soil ecosystems. Soils form composite phenotypes where inorganic matter is enveloped into the physiology of a whole community. As organisms feed and migrate through soils they physically displace materials, an ecological process called bioturbation. This aerates soils and stimulates heterotrophic growth and production. Soil microorganisms are influenced by and feed back into the trophic dynamics of the ecosystem. No single axis of causality can be discerned to segregate the biological from geomorphological systems in soils. Paleoecological studies of soils places the origin for bioturbation to a time before the Cambrian period. Other events, such as the evolution of trees and the colonization of land in the Devonian period played a significant role in the early development of ecological trophism in soils.

Biogeochemistry and Climate

Ecologists study and measure nutrient budgets to understand how these materials are regulated, flow, and recycled through the environment. This research has led to an understanding that there is global feedback between ecosystems and the physical parameters of this planet, including minerals, soil, pH, ions, water, and atmospheric gases. Six major elements (hydrogen, carbon, nitrogen, oxygen, sulfur, and phosphorus; H, C, N, O, S, and P) form the constitution of all biological macromolecules and feed into the Earth's geochemical processes. From the smallest scale of biology, the combined effect of billions upon billions of ecological processes amplify and ultimately regulate the biogeochemical cycles of the Earth. Understanding the relations and cycles mediated between these elements and their ecological pathways has significant bearing toward understanding global biogeochemistry.

The ecology of global carbon budgets gives one example of the linkage between biodiversity and biogeochemistry. It is estimated that the Earth's oceans hold 40,000 gigatonnes (Gt) of carbon, that vegetation and soil hold 2070 Gt, and that fossil fuel emissions are 6.3 Gt carbon per year. There have been major restructurings in these global carbon budgets during the Earth's history, regulated to a large extent by the ecology of the land. For example, through the early-mid Eocene volcanic outgassing, the oxidation of methane stored in wetlands, and seafloor gases increased atmospheric CO_2 (carbon dioxide) concentrations to levels as high as 3500 ppm.

In the Oligocene, from twenty-five to thirty-two million years ago, there was another significant restructuring of the global carbon cycle as grasses evolved a new mechanism of photosynthesis, C_4 photosynthesis, and expanded their ranges. This new pathway evolved in response to the drop in atmospheric CO_2 concentrations below 550 ppm. The relative abundance and distribution of biodiversity alters the dynamics between organisms and their environment such that ecosystems can be both cause and effect in relation to climate change. Human-driven modifications to the planet's ecosystems (e.g., disturbance, biodiversity loss, agriculture) contributes to rising atmospheric greenhouse gas levels. Transformation of the global carbon cycle in the next century is projected to raise planetary temperatures, lead to more extreme fluctuations in weather, alter species distributions, and increase extinction rates. The effect of global warming is already being registered in melting glaciers, melting mountain ice caps, and rising sea levels. Consequently, species distributions are changing along waterfronts and in continental areas where migration patterns and breeding grounds are tracking the prevailing shifts in climate. Large sections of permafrost are also melting to create a new mosaic of flooded areas having increased rates of soil decomposition activity that raises methane (CH_4) emissions. There is concern over increases in atmospheric methane in the context of the global carbon cycle, because methane is a greenhouse gas that is 23 times more effective at absorbing long-wave radiation than CO_2 on a 100-year time scale. Hence, there is a relationship between global warming, decomposition and respiration in soils and wetlands producing significant climate feedbacks and globally altered biogeochemical cycles.

History

Early Beginnings

Ecology has a complex origin, due in large part to its interdisciplinary nature. Ancient Greek philosophers such as Hippocrates and Aristotle were among the first to record observations on natural history. However, they viewed life in terms of essentialism, where species were conceptualized as static unchanging things while varieties were seen as aberrations of an idealized type. This contrasts against the modern understanding of ecological theory where varieties are viewed as the real phenomena of interest and having a role in the origins of adaptations by means of natural selection. Early conceptions of ecology, such as a balance and regulation in nature can be traced to Herodotus (died c. 425 BC), who described one of the earliest accounts of mutualism in his observation of "natural dentistry". Basking Nile crocodiles, he noted, would open their mouths to give sandpipers safe access to pluck leeches out, giving nutrition to the sandpiper and oral hygiene for the crocodile. Aristotle was an early influence on the philosophical development of ecology. He and his student Theophrastus made extensive observations on plant and animal migrations, biogeography, physiology, and on their behaviour, giving an early analogue to the modern concept of an ecological niche.

Ecological concepts such as food chains, population regulation, and productivity were first developed in the 1700s, through the published works of microscopist Antoni van Leeuwenhoek (1632–1723) and botanist Richard Bradley (1688?–1732). Biogeographer Alexander von Humboldt (1769–1859) was an early pioneer in ecological thinking and was among the first to recognize ecological gradients, where species are replaced or altered in form along environmental gradients, such as a cline forming along a rise in elevation. Humboldt drew inspiration from Isaac Newton as he developed a form of "terrestrial physics". In Newtonian fashion, he brought a scientific exactitude for measurement into natural history and even alluded to concepts that are the foundation of a modern ecological law on species-to-area relationships. Natural historians, such as Humboldt, James Hutton, and Jean-Baptiste Lamarck (among others) laid the foundations of the modern ecological sciences. The term "ecology" (German: *Oekologie, Ökologie*) is of a more recent origin and was first coined by the German biologist Ernst Haeckel in his book *Generelle Morphologie der Organismen* (1866). Haeckel was a zoologist, artist, writer, and later in life a professor of comparative anatomy.

By ecology, we mean the whole science of the relations of the organism to the environment including, in the broad sense, all the "conditions of existence. "Thus the theory of evolution explains the housekeeping relations of organisms mechanistically as the necessary consequences of effectual causes and so forms the monistic groundwork of ecology.

Ernst Haeckel (1866)

Ernst Haeckel (left) and Eugenius Warming (right), two founders of ecology

Opinions differ on who was the founder of modern ecological theory. Some mark Haeckel's definition as the beginning; others say it was Eugenius Warming with the writing of Oecology of Plants: An Introduction to the Study of Plant Communities (1895), or Carl Linnaeus' principles on the economy of nature that matured in the early 18th century. Linnaeus founded an early branch of ecology that he called the economy of nature. His works influenced Charles Darwin, who adopted Linnaeus' phrase on the *economy or polity of nature* in *The Origin of Species*. Linnaeus was the first to frame the balance of nature as a testable hypothesis. Haeckel, who admired Darwin's work, defined ecology in reference to the economy of nature, which has led some to question whether ecology and the economy of nature are synonymous.

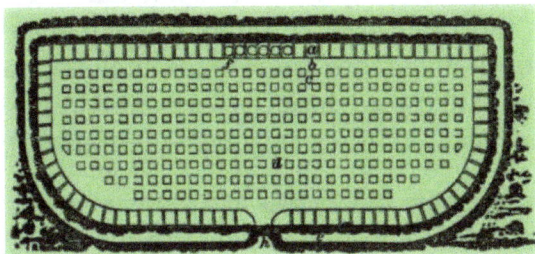

The layout of the first ecological experiment, carried out in a grass garden at Woburn Abbey in 1816, was noted by Charles Darwin in *The Origin of Species*. The experiment studied the performance of different mixtures of species planted in different kinds of soils.

From Aristotle until Darwin, the natural world was predominantly considered static and unchanging. Prior to *The Origin of Species*, there was little appreciation or understanding of the dynamic and reciprocal relations between organisms, their adaptations, and the environment. An exception is the 1789 publication *Natural History of Selborne* by Gilbert White (1720–1793), considered by some to be one of the earliest texts on ecology. While Charles Darwin is mainly noted for his treatise on evolution, he was one of the founders of soil ecology, and he made note of the first ecological experiment in *The Origin of Species*. Evolutionary theory changed the way that researchers approached the ecological sciences.

Nowhere can one see more clearly illustrated what may be called the sensibility of such an organic complex,--expressed by the fact that whatever affects any species belonging to it, must speedily have its influence of some sort upon the whole assemblage. He will thus be made to see the impossibility of studying any form completely, out of relation to the other forms,--the necessity for taking a comprehensive survey of the whole as a condition to a satisfactory understanding of any part.

Stephen Forbes (1887)

Since 1900

Modern ecology is a young science that first attracted substantial scientific attention toward the end of the 19th century (around the same time that evolutionary studies were gaining scientific interest). Notable scientist Ellen Swallow Richards may have first introduced the term "oekology" (which eventually morphed into home economics) in the U.S. as early 1892.

In the early 20th century, ecology transitioned from a more descriptive form of natural history to a more analytical form of *scientific natural history*. Frederic Clements published the first American ecology book in 1905, presenting the idea of plant communities as a superorganism. This publication launched a debate between ecological holism and individualism that lasted until the 1970s. Clements' superorganism concept proposed that ecosystems progress through regular and determined stages of seral development that are analogous to the developmental stages of an organism. The Clementsian paradigm was challenged by Henry Gleason, who stated that ecological communities develop from the unique and coincidental association of individual organisms. This perceptual shift placed the focus back onto the life histories of individual organisms and how this relates to the development of community associations.

The Clementsian superorganism theory was an overextended application of an idealistic form of holism. The term "holism" was coined in 1926 by Jan Christiaan Smuts, a South African general and polarizing historical figure who was inspired by Clements' superorganism concept.[c] Around the same time, Charles Elton pioneered the concept of food chains in his classical book *Animal Ecology*. Elton defined ecological relations using concepts of food chains, food cycles, and food size, and described numerical relations among different functional groups and their relative abundance. Elton's 'food cycle' was replaced by 'food web' in a subsequent ecological text. Alfred J. Lotka brought in many theoretical concepts applying thermodynamic principles to ecology.

In 1942, Raymond Lindeman wrote a landmark paper on the trophic dynamics of ecology, which was published posthumously after initially being rejected for its theoretical emphasis. Trophic dynamics became the foundation for much of the work to follow on energy and material flow through ecosystems. Robert MacArthur advanced mathematical theory, predictions, and tests in ecology in the 1950s, which inspired a resurgent school of theoretical mathematical ecologists. Ecology also has developed through contributions from other nations, including Russia's Vladimir Vernadsky and his founding of the biosphere concept in the 1920s and Japan's Kinji Imanishi and his concepts of harmony in nature and habitat segregation in the 1950s. Scientific recognition of contributions to ecology from non-English-speaking cultures is hampered by language and translation barriers.

This whole chain of poisoning, then, seems to rest on a base of minute plants which must have been the original concentrators. But what of the opposite end of the food chain—the human being who, in probable ignorance of all this sequence of events, has rigged his fishing tackle, caught a string of fish from the waters of Clear Lake, and taken them home to fry for his supper?

Rachel Carson (1962)

Ecology surged in popular and scientific interest during the 1960–1970s environmental movement. There are strong historical and scientific ties between ecology, environmental management, and protection. The historical emphasis and poetic naturalistic writings advocating the protection of wild places by notable ecologists in the history of conservation biology, such as Aldo Leopold and Arthur Tansley, have been seen as far removed from urban centres where, it is claimed, the concentration of pollution and environmental degradation is located. Palamar (2008) notes an overshadowing by mainstream environmentalism of pioneering women in the early 1900s who fought for urban health ecology (then called euthenics) and brought about changes in environmental legislation. Women such as Ellen Swallow Richards and Julia Lathrop, among others, were precursors to the more popularized environmental movements after the 1950s.

In 1962, marine biologist and ecologist Rachel Carson's book *Silent Spring* helped to mobilize the environmental movement by alerting the public to toxic pesticides, such as DDT, bioaccumulating in the environment. Carson used ecological science to link the release of environmental toxins to human and ecosystem health. Since then, ecologists have worked to bridge their understanding of the degradation of the planet's ecosystems with environmental politics, law, restoration, and natural resources management.

Population Ecology

Population ecology or autecology is a sub-field of ecology that deals with the dynamics of species populations and how these populations interact with the environment. It is the study of how the population sizes of species change over time and space. The term population ecology is often used interchangeably with population biology or population dynamics.

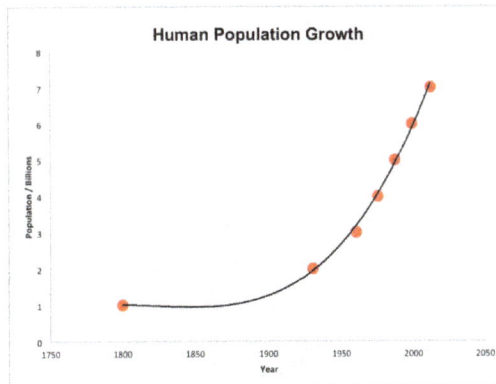

The human population is growing at an exponential rate and is affecting the populations of other species in return. Chemical pollution, deforestation, and irrigation are examples of means by which humans may influence the population ecology of other species. As the human population increases, its effect on the populations of other species may also increase.

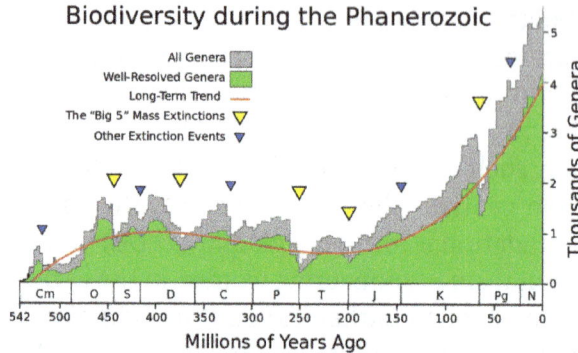

Biodiversity during the Phanerozoic

Populations cannot grow indefinitely. Population ecology involves studying factors that affect population growth and survival. Mass extinctions are examples of factors that have radically reduced populations' sizes and populations' survivability. The survivability of populations is critical to maintaining high levels of biodiversity on Earth.

The development of population ecology owes much to demography and actuarial life tables. Population ecology is important in conservation biology, especially in the development of population viability analysis (PVA) which makes it possible to predict the long-term probability of a species persisting in a given habitat patch. Although population ecology is a subfield of biology, it provides interesting problems for mathematicians and statisticians who work in population dynamics.

Fundamentals

Terms used to describe natural groups of individuals in ecological studies

Term	Definition
Species population	All individuals of a species.
Metapopulation	A set of spatially disjunct populations, among which there is some immigration.
Population	A group of conspecific individuals that is demographically, genetically, or spatially disjunct from other groups of individuals.
Aggregation	A spatially clustered group of individuals.
Deme	A group of individuals more genetically similar to each other than to other individuals, usually with some degree of spatial isolation as well.
Local population	A group of individuals within an investigator-delimited area smaller than the geographic range of the species and often within a population (as defined above). A local population could be a disjunct population as well.
Subpopulation	An arbitrary spatially delimited subset of individuals from within a population (as defined above).

The most fundamental law of population ecology is Thomas Malthus' exponential law of population growth.

A population will grow (or decline) exponentially as long as the environment experienced by all individuals in the population remains constant.

Thomas Robert Malthus

This principle in population ecology provides the basis for formulating predictive theories and tests that follow:

Simplified population models usually start with four key variables (four demographic processes) including death, birth, immigration, and emigration. Mathematical models used to calculate changes in population demographics and evolution hold the assumption (or null hypothesis) of no external influence. Models can be more mathematically complex where "...several competing hypotheses are simultaneously confronted with the data." For example, in a closed system where immigration and emigration does not take place, the rate of change in the number of individuals in a population can be described as:

$$\frac{dN}{dT} = B - D = bN - dN = (b - d)N = rN,$$

where N is the total number of individuals in the population, B is the raw number of births, D is the raw number of deaths, b and d are the per capita rates of birth and death respectively, and r is the per capita average number of surviving offspring each individual has. This formula can be read as the rate of change in the population (dN/dT) is equal to births minus deaths (B - D).

Using these techniques, Malthus' population principle of growth was later transformed into a mathematical model known as the logistic equation:

$$\frac{dN}{dT} = aN\left(1 - \frac{N}{K}\right),$$

where N is the biomass density, a is the maximum per-capita rate of change, and K is the carrying capacity of the population. The formula can be read as follows: the rate of change in the population (*dN/dT*) is equal to growth (*aN*) that is limited by carrying capacity *(1-N/K)*. From these basic mathematical principles the discipline of population ecology expands into a field of investigation that queries the demographics of real populations and tests these results against the statistical models. The field of population ecology often uses data on life history and matrix algebra to develop projection matrices on fecundity and survivorship. This information is used for managing wildlife stocks and setting harvest quotas

Geometric Populations

Operophtera brumata (Winter moth) populations are geometric.

The population model below can be manipulated to mathematically infer certain properties of geometric populations. A population with a size that increases geometrically is a population where generations of reproduction do not overlap. In each generation there is an effective population size denoted as N_e which constitutes the number of individuals in the population that are able to reproduce *and* will reproduce in any reproductive generation in concern. In the population model below it is assumed that N is the effective population size.

Assumption 01: $N_e = N$

$$N_{t+1} = N_t + B_t + I_t - D_t - E_t$$

Term	Definition
N_{t+1}	Population size in the generation after generation $_t$. This may be the current generation or the next (upcoming) generation depending on the situation in which the population model is used.
N_t	Population size in generation $_t$.
B_t	Sum (Σ) of births in the population between generations $_t$ and $_{t+1}$. Also known as raw birth rate.

Term	Definition
I_t	Sum (Σ) of immigrants moving into the population between generations $_t$ and $_{t+1}$. Also known as raw immigration rate.
D_t	Sum (Σ) of deaths in the population between generations $_t$ and $_{t+1}$. Also known as raw death rate.
E_t	Sum (Σ) of emigrants moving out of the population between generations $_t$ and $_{t+1}$. Also known as raw emigration rate.

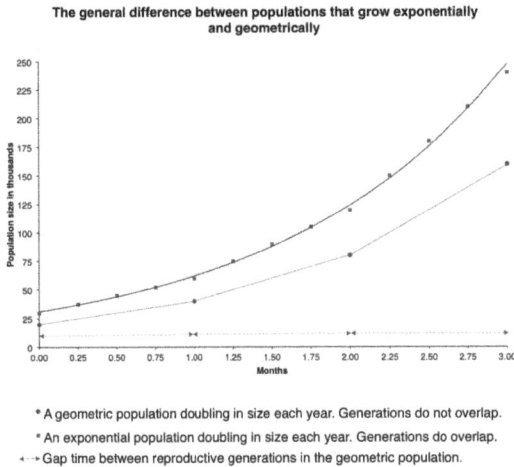

The general difference between populations that grow exponentially and geometrically

* A geometric population doubling in size each year. Generations do not overlap.
* An exponential population doubling in size each year. Generations do overlap.
←--→ Gap time between reproductive generations in the geometric population.

The general difference between populations that grow exponentially and geometrically. Geometric populations grow in reproductive generations between intervals of abstinence from reproduction. Exponential populations grow without designated periods for reproduction. Reproduction is a continuous process and generations of reproduction overlap. This graph illustrates two hypothetical populations - one population growing periodically (and therefore geometrically) and the other population growing continuously (and therefore exponentially). The populations in the graph have a doubling time of 1 year. The populations in the graph are hypothetical. In reality, the doubling times differ between populations.

Assumption 02: There is no migration to or from the population (N)

$$I_t = E_t = 0$$

$$N_{t+1} = N_t + B_t - D_t$$

The raw birth and death rates are related to the per capita birth and death rates:

$$B_t = b_t \times N_t$$

$$D_t = d_t \times N_t$$

$$b_t = B_t / N_t$$

$$d_t = D_t / N_t$$

Term	Definition
b_t	Per capita birth rate.
d_t	Per capita death rate.

Therefore:

$$N_{t+1} = N_t + (b_t \times N_t) - (d_t \times N_t)$$

Assumption 03: b_t and d_t are constant (i.e. they don't change each generation).

$$N_{t+1} = N_t + (bN_t) - (dN_t)$$

Term	Definition
b	Constant per capita birth rate.
d	Constant per capita death rate.

Take the term N_t out of the brackets.

$$N_{t+1} = N_t + (b - d)N_t$$

$$b - d = R$$

Term	Definition
R	Geometric rate of increase.

$$N_{t+1} = N_t + RN_t$$

$$N_{t+1} = (N_t + RN_t)$$

Take the term N_t out of the brackets again.

$$N_{t+1} = (1 + R)N_t$$

$$1 + R = \lambda$$

Term	Definition
λ	Finite rate of increase.

$$N_{t+1} = \lambda N_t$$

At $_{t+1}$ $N_{t+1} = \lambda N_t$

At $_{t+2}$ $N_{t+2} = \lambda N_{t+1} = \lambda\lambda N_t = \lambda^2 N_t$

At $_{t+3}$ $N_{t+3} = \lambda N_{t+2} = \lambda\lambda N_{t+1} = \lambda\lambda\lambda N_t = \lambda^3 N_t$

At $_{t+4}$ $N_{t+4} = \lambda N_{t+3} = \lambda\lambda N_{t+2} = \lambda\lambda\lambda N_{t+1} = \lambda\lambda\lambda\lambda N_t = \lambda^4 N_t$

At $_{t+5}$ $N_{t+5} = \lambda N_{t+4} = \lambda\lambda N_{t+3} = \lambda\lambda\lambda N_{t+2} = \lambda\lambda\lambda\lambda N_{t+1} = \lambda\lambda\lambda\lambda\lambda N_t = \lambda^5 N_t$

Therefore:

$$N_{t+1} = \lambda^t N_t$$

Term	Definition
λ^t	Finite rate of increase raised to the power of the number of generations (e.g. for $_{t+2}$ [two generations] $\rightarrow \lambda^2$, for $_{t+1}$ [one generation] $\rightarrow \lambda^1 = \lambda$, and for $_t$ [before any generations - at time zero] $\rightarrow \lambda^0 = 1$

Doubling Time of Geometric Populations

G. stearothermophilus has a shorter doubling time (td) than E. coli and N. meningitidis. Growth rates of 2 bacterial species will differ by unexpected orders of magnitude if the doubling times of the 2 species differ by even as little as 10 minutes. In eukaryotes such as animals, fungi, plants, and protists, doubling times are much longer than in bacteria. This reduces the growth rates of eukaryotes in comparison to Bacteria. *G. stearothermophilus*, *E. coli*, and *N. meningitidis* have 20 minute, 30 minute, and 40 minute doubling times under optimal conditions respectively. If bacterial populations could grow indefinitely (which they do not) then the number of bacteria in each species would approach infinity (∞). However, the percentage of *G. stearothermophilus* bacteria out of all the bacteria would approach 100% whilst the percentage of *E. coli* and *N. meningitidis* combined out of all the bacteria would approach 0%. This graph is a simulation of this hypothetical scenario. In reality, bacterial populations do not grow indefinitely in size and the 3 species require different optimal conditions to bring their doubling times to minima.

Time in minutes	% that is *E. coli*
30	29.6%
60	26.7%
90	21.6%
120	18.2%
$\rightarrow\infty$	0.00%

Time in minutes	% that is *N. meningitidis*
30	25.9%
60	20.0%

90	13.5%
120	9.10%
→∞	0.00%

Disclaimer: Bacterial populations are exponential (or, more correctly, logistic) instead of geometric. Nevertheless, doubling times are applicable to both types of populations.

The doubling time of a population is the time required for the population to grow to twice its size. We can calculate the doubling time of a geometric population using the equation: $N_{t+1} = \lambda^t N_t$ by exploiting our knowledge of the fact that the population (N) is twice its size (2N) after the doubling time.

$$2N_{td} = \lambda^t_d \times N_t$$

Term	Definition
t_d	Doubling time.

$$\lambda^t_d = 2N_{td} / N_t$$

$$\lambda^t_d = 2$$

The doubling time can be found by taking logarithms. For instance:

$$t_d \times \log_2(\lambda) = \log_2(2)$$

$$\log_2(2) = 1$$

$$t_d \times \log_2(\lambda) = 1$$

$$t_d = 1 / \log_2(\lambda)$$

Or:

$$t_d \times \ln(\lambda) = \ln(2)$$

$$t_d = \ln(2) / \ln(\lambda)$$

$$t_d = 0.693... / \ln(\lambda)$$

Therefore:

$$t_d = 1 / \log_2(\lambda) = 0.693... / \ln(\lambda)$$

Half-life of Geometric Populations

The half-life of a population is the time taken for the population to decline to half its size. We can calculate the half-life of a geometric population using the equation: N_{t+1}

$= \lambda^t N_t$ by exploiting our knowledge of the fact that the population (N) is half its size (0.5N) after a half-life.

$0.5N_{t1/2} = \lambda^t_{1/2} \times N_t$

Term	Definition
$t_{1/2}$	Half-life.

$\lambda^t_{1/2} = 0.5N_{t1/2} / N_t$

$\lambda^t_{1/2} = 0.5$

The half-life can be calculated by taking logarithms.

$t_{1/2} = 1 / \log_{0.5}(\lambda) = \ln(0.5) / \ln(\lambda)$

Geometric (R) and finite (λ) growth constants

Geometric (R) growth constant

$R = b - d$

$N_{t+1} = N_t + RN_t$

$N_{t+1} - N_t = RN_t$

$N_{t+1} - N_t = \Delta N$

Term	Definition
ΔN	Change in population size between two generations (between generation $_{t+1}$ and $_t$).

$\Delta N = RN_t$

$\Delta N/N_t = R$

Finite (λ) growth constant

$1 + R = \lambda$

$N_{t+1} = \lambda N_t$

$\lambda = N_{t+1} / N_t$

Mathematical Relationship between Geometric and Exponential Populations

In geometric populations, R and λ represent growth constants. In exponential populations however, the intrinsic growth rate, also known as intrinsic rate of increase

(r) is the relevant growth constant. Since generations of reproduction in a geometric population do not overlap (e.g. reproduce once a year) but do in an exponential population, geometric and exponential populations are usually considered to be mutually exclusive. However, geometric constants and exponential constants share the mathematical relationship below.

The growth equation for exponential populations is

$$N_t = N_o e^{rt}$$

Term	Definition
e	Euler's number - A universal constant often applicable in exponential equations.
r	intrinsic growth rate - also known as intrinsic rate of increase.

Leonhard Euler was the mathematician who established the universal constant 2.71828... also known as Euler's number or e.

Assumption: N_t *(of a geometric population)* = N_t *(of an exponential population)*.

Therefore:

$$N_o e^{rt} = N_o \lambda^t$$

N_o cancels on both sides.

$$N_o e^{rt} / N_o = \lambda^t$$

$$e^{rt} = \lambda^t$$

Take the natural logarithms of the equation. Using natural logarithms instead of base 10 or base 2 logarithms simplifies the final equation as $\ln(e) = 1$.

$$rt \times \ln(e) = t \times \ln(\lambda)$$

Term	Definition
ln	natural logarithm - in other words $\ln(y) = \log_e(y) = x$ = the power (x) that e needs to be raised to (e^x) to give the answer y. In this case, e^1 = e therefore $\ln(e) = 1$.

$rt \times 1 = t \times \ln(\lambda)$

$rt = t \times \ln(\lambda)$

t cancels on both sides.

$rt / t = \ln(\lambda)$

The results:

$r = \ln(\lambda)$

and

$e^r = \lambda$

r/K selection

At its most elementary level, interspecific competition involves two species utilizing a similar resource. It rapidly gets more complicated, but stripping the phenomenon of all its complications, this is the basic principle: two consumers consuming the same resource.

An important concept in population ecology is the r/K selection theory. The first variable is r (the intrinsic rate of natural increase in population size, density independent) and the second variable is K (the carrying capacity of a population, density dependent). An r-selected species (e.g., many kinds of insects, such as aphids) is one that has high rates of fecundity, low levels of parental investment in the young, and high rates of mortality before individuals reach maturity. Evolution favors productivity in r-selected species. In contrast, a K-selected species (such as humans) has low rates of fecundity, high levels of parental investment in the young, and low rates of mortality as individuals mature. Evolution in K-selected species favors efficiency in the conversion of more resources into fewer offspring.

Metapopulation

Populations are also studied and conceptualized through the "metapopulation" concept. The metapopulation concept was introduced in 1969:

"as a population of populations which go extinct locally and recolonize."

Metapopulation ecology is a simplified model of the landscape into patches of varying levels of quality. Patches are either occupied or they are not. Migrants moving among the patches are structured into metapopulations either as sources or sinks. Source patches are productive sites that generate a seasonal supply of migrants to other patch locations. Sink patches are unproductive sites that only receive migrants. In metapopulation terminology there are emigrants (individuals that leave a patch) and immigrants (individuals that move into a patch). Metapopulation models examine patch dynamics over time to answer questions about spatial and demographic ecology. An important concept in metapopulation ecology is the rescue effect, where small patches of lower quality (i.e., sinks) are maintained by a seasonal influx of new immigrants. Metapopulation structure evolves from year to year, where some patches are sinks, such as dry years, and become sources when conditions are more favorable. Ecologists utilize a mixture of computer models and field studies to explain metapopulation structure.

History

The older term, autecology (*auto*, "self"; "household"; and logos, "knowledge"), refers to roughly the same field of study as population ecology. It derives from the division of ecology into autecology—the study of individual species in relation to the environment—and synecology—the study of groups of organisms in relation to the environment—or community ecology. Odum (1959, p. 8) considered that synecology should be divided into population ecology, community ecology, and ecosystem ecology, defining autecology as essentially "species ecology." However, for some time biologists have recognized that the more significant level of organization of a species is a population, because at this level the species gene pool is most coherent. In fact, Odum regarded "autecology" as no longer a "present tendency" in ecology (i.e., an archaic term), although included "species ecology"—studies emphasizing life history and behavior as adaptations to the environment of individual organisms or species—as one of four subdivisions of ecology.

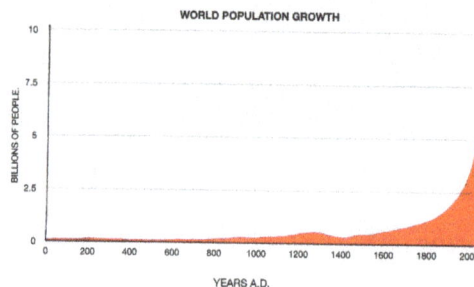

WORLD POPULATION GROWTH

Restoration Ecology

Restoration ecology emerged as a separate field in ecology in the 1980s. It is the scientific study supporting the practice of ecological restoration, which is the practice of renewing and restoring degraded, damaged, or destroyed ecosystems and habitats in the environment by active human intervention and action. Restoration ecology is therefore commonly used for the academic study of the process, whereas ecological restoration is commonly used for the actual project or process by restoration practitioners.

Recently constructed wetland regeneration in Australia, on a site previously used for agriculture

Rehabilitation of a portion of Johnson Creek, to restore bioswale and flood control functions of the land which had long been converted to pasture for cow grazing. The horizontal logs can float, but are anchored by the posts. Just-planted trees will eventually stabilize the soil. The fallen trees with roots jutting into the stream are intended to enhance wildlife habitat. The meandering of the stream is enhanced here by a factor of about three times, perhaps to its original course.

Definition

The Society for Ecological Restoration defines *"ecological restoration"* as an "intentional activity that initiates or accelerates the recovery of an ecosystem with respect to its health, integrity and sustainability". The practice of ecological restoration includes wide scope of projects such as erosion control, reforestation, usage of genetically local native species, removal of non-native species and weeds, revegetation of disturbed areas, daylighting streams, reintroduction of native species, as well as habitat and range improvement for targeted species.

E. O. Wilson, a biologist states that: "Here is the means to end the great extinction spasm. The next century will, I believe, be the era of restoration in ecology."

Restoration ecology is the scientific study of ecological restoration.

History

Land managers, laypeople, and stewards have been practicing ecological restoration or ecological management for many hundreds, if not thousands of years, yet the scientific field of "restoration ecology" was not first formally identified and coined until the late 1980s, by John Aber and William Jordan when they were at the University of Wisconsin-Madison. They held the first international meetings on this topic in Madison during which attendees visited the University of Wisconsin's Arboretum—the oldest restoration ecology project made famous by Professor Aldo Leopold. The study of restoration ecology has only become a robust and independent scientific discipline over the last two decades, and the commercial applications of ecological restoration have tremendously increased in recent years.

Restoration Needs

There is consensus in the scientific community that the current environmental degradation and destruction of many of the Earth's biota is considerable and is taking place on a "catastrophically short timescale". Estimates of the current extinction rate is 1000 to 10,000 times more than the normal rate. For many people biological diversity, (biodiversity) has an intrinsic value that humans have a responsibility towards other living things, and an obligation to future generations.

On a more anthropocentric level, natural ecosystems provide human society with food, fuel and timber. Fundamentally, ecosystem services involve the purification of air and water, detoxification and decomposition of wastes, regulation of climate, regeneration of soil fertility and pollination of crops. Such processes have been estimated to be worth trillions of dollars annually.

Habitat loss is the leading cause of both species extinctions and ecosystem service decline. The two ways to reverse this trend of habitat loss are conservation of currently viable habitat and restoration of degraded habitats.

Conservation Biology and Restoration Ecology

Restoration ecology may be viewed as a sub-discipline of conservation biology, the scientific study of how to protect and restore biodiversity, and restoration a part of the resulting conservation movement.

Conservation Vs. Restoration

The fundamental difference between restoration and other conservation efforts is analogous to the difference between disease prevention and treatment. Conservation attempts to maintain and protect existing habitat and biodiversity, whereas restoration attempts to reverse existing environmental degradation and population declines. Targeted human intervention is used to promote habitat, biodiversity recovery and associated gains.

The possibility of restoration, however, does not provide an excuse for converting extremely valuable "pristine" habitat into other uses: as in medicine, it better to prevent than to treat. "Treatment" is generally less effective and more expensive than prevention, and "treatment" cannot always restore the condition before the "injury": some habitat and biodiversity losses are permanent.

Focuses

Though restoration ecologists and other conservation biologists generally agree that habitat is the most important locus of biodiversity protection, the disciplines themselves have different focuses. Conservation biology as an academic discipline is rooted in population biology. Because of that, it is generally organized at the genetic level, looking at specific species populations (i.e. endangered species). Restoration ecology is organized at the community level, looking at specific ecosystems.

Because it is organized by species, conservation biology often emphasizes vertebrate animals because of their salience and popularity, whereas restoration ecology emphasizes plants because of their importance within ecosystems: ecosystem restoration is botanically based. Since soils define the foundation of any functional terrestrial system, restoration ecology's ecosystem-level focus also results in greater emphasis on the role of soil's physical and microbial processes.

Modes of Inquiry

Conservation biology's focus on rare or endangered species limit the number of manipulative studies that can be performed. As a consequence, conservation studies tend to be descriptive, comparative and unreplicable. However, the highly manipulative nature of restoration ecology allows the researcher to test the hypotheses vigorously. Restorative activity often reflects an experimental test of what limits populations.

Theoretical Foundations

Restoration ecology draws on a wide range of ecological concepts.

Disturbance

Disturbance is a change of environmental conditions, which interferes with the functioning of a biological system. Disturbance, at a variety of spatial and temporal scales is a natural, and even essential, component of many communities.

Humans have had limited "natural" impacts on ecosystems for as long as humans have existed, however, the severity and scope of our influences has accelerated in the last few centuries. Understanding and minimizing the differences between modern anthropogenic and "natural" disturbances is crucial to restoration ecology. For example, new forestry techniques that better imitate historical disturbances are now being implemented.

In addition, restoring a fully sustainable ecosystem often involves studying and attempting to restore a natural disturbance regime (e.g., fire ecology).

Succession

Ecological succession is the process by which the component species of a community changes over time. Following a disturbance, an ecosystem generally progresses from a simple level of organization (i.e. few dominant species) to a more complex community (i.e. many interdependent species) over few generations. Depending on the severity of the disturbance, restoration often consists of initiating, assisting or accelerating ecological successional processes.

In many ecosystems, communities tend to recover following mild to moderate natural and anthropogenic disturbances. Restoration in these systems involves hastening natural successional trajectories. However, a system that has experienced a more severe disturbance (i.e. physical or chemical alteration of the environment) may require intensive restorative efforts to recreate environmental conditions that favor natural successional processes. This ability to recover is called resilience.

Fragmentation

Habitat fragmentation is the emergence of spatial discontinuities in a biological system. Through land use changes (e.g. agriculture) and "natural" disturbance, ecosystems are broken up into smaller parts. Small fragments of habitat can support only small populations and small populations are more vulnerable to extinction. Furthermore, fragmenting ecosystems decreases interior habitat. Habitat along the edge of a fragment has a different range of environmental conditions and therefore supports different species than the interior. Fragmentation effectively reduces interior habitat and may lead to the extinction of those species which require interior habitat. Restorative projects can increase the effective size of a habitat by simply adding area or by planting habitat corridors that link and fill in the gap between two isolated fragments. Reversing the effects of fragmentation and increasing habitat connectivity are the central goals of restoration ecology.

Ecosystem Function

Ecosystem function describes the foundational processes of natural systems, including nutrient cycles and energy fluxes. These processes are the most basic and essential components of ecosystems. An understanding of the full complexity and intricacies of these cycles is necessary to address any ecological processes that may be degraded. A functional ecosystem, that is completely self-perpetuating (no management required), is the ultimate goal of restorative efforts. Since, these ecosystem functions are emergent properties of the system as a whole, monitoring and management are crucial for the long-term stability of an ecosystem.

Evolving Concepts

Restoration ecology, because of its highly physical nature, is an ideal testing ground for an emerging community's ecological principles (Bradshaw 1987). Likewise, there are emerging concepts of inventing new and successful restoration technologies, performance standards, time frames, local genetics, and society's relationship to restoration ecology, and new ethical and religious possibilities, as future topics of discussion and debate.

Assembly

Community assembly "is a framework that can unify virtually all of (community) ecology under a single conceptual umbrella". Community assembly theory attempts to explain the existence of environmentally similar sites with differing assemblages of species. It assumes that species have similar niche requirements, so that community formation is a product of random fluctuations from a common species pool. Essentially, if all species are fairly ecologically equivalent then random variation in colonization, migration and extinction rates between species, drive differences in species composition between sites with comparable environmental conditions.

Stable States

Alternative stable states are discrete species compositional possibilities that may exist within a community. According to assembly theory, differences in species colonization, interspecific interactions and community establishment may result in distinct community species equilibria. A community has numerous possible compositional equilibria that are dependent on the initial assembly. That is, random fluctuations lead to a particular initial community assembly, which affects successional trajectories and the eventual species composition equilibrium.

Multiple stable states is a specific theoretical concept, where all species have equal access to a community (i.e., equal dispersal potential) and differences between communities arise simply because of the timing of each species' colonization.

These concepts are central to restoration ecology; restoring a community not only involves manipulating the timing and structure of the initial species composition, but also working towards a single desired stable state. In fact, a degraded ecosystem may be viewed as an alternative stable state under the altered environmental conditions.

Ontogeny

The ecology of ontogeny is the study of how ecological relationships change over the lifetime of an individual. Organisms require different environmental conditions during different stages of their life-cycle. For immobile organisms (e.g. plants), the conditions necessary for germination and establishment may be different from those of the adult stage. As an ecosystem is altered by anthropogenic processes the range of environmental variables may also be altered. A degraded ecosystem may not include the environmental conditions necessary for a particular stage of an organism's development. If a self-sustaining, functional ecosystem must contain environmental conditions for the perpetual reproduction of its species, restorative efforts must address the needs of organisms throughout their development.

Application of Theory

Restoration is defined as the application of ecological theory to ecological restoration. However, for many reasons, this can be a challenging prospect. Here are a few examples of theory informing practice.

Soil heterogeneity Effects on Community Heterogeneity

Spatial heterogeneity of resources can influence plant community composition, diversity and assembly trajectory. Baer et al. (2005) manipulated soil resource heterogeneity in a tallgrass prairie restoration project. They found increasing resource heterogeneity, which on its own was insufficient to insure species diversity in situations where one species may dominate across the range of resource levels. Their findings were consistent with the theory regarding the role of ecological filters on community assembly. The establishment of a single species, best adapted to the physical and biological conditions can play an inordinately important role in determining the community structure.

Invasion, Competitive Dominance and Resource Use

"The dynamics of invasive species may depend on their abilities to compete for resources and exploit disturbances relative to the abilities of native species". Seabloom et al. (2003) tested this concept and its implications in a California grassland restoration context. They found that the native grass species were able to successfully compete with invasive exotics, therefore, the possibility exists of restoring an original native grassland ecosystem.

Successional Trajectories

Progress along a desired successional pathway may be difficult if multiple stable states exist. Looking over 40 years of wetland restoration data Klotzi and Gootjans (2001) argue that unexpected and undesired vegetation assemblies "may indicate that environmental conditions are not suitable for target communities". Succession may move in unpredicted directions, but constricting environmental conditions within a narrow range may rein in the possible successional trajectories and increase the likelihood of a desired outcome.

Natural Capital Committee's Recommendation for a 25-year plan

The UK Natural Capital Committee (NCC) made a recommendation in its second State of Natural Capital report published in March 2014 that in order to meet the Government's goal of being the first generation to leave the environment in a better state than it was inherited, a long-term 25-year plan was needed to maintain and improve England's natural capital. The UK Government has not yet responded to this recommendation.

The Secretary of State for the UK's Department for Environment, Food and Rural Affairs, Owen Paterson, described his ambition for the natural environment and how the work of the Committee fits into this at an NCC event in November 2012: "I do not, however, just want to maintain our natural assets; I want to improve them. I want us to derive the greatest possible benefit from them, while ensuring that they are available for generations to come. This is what the NCC's innovative work is geared towards".

Ecosystem Restoration

Ecosystem restoration for the superb parrot on an abandoned railway line in Australia

According to the Society for Ecological Restoration, ecosystem restoration is the return of a damaged ecological system to a stable, healthy, and sustainable state, often together with associated ecosystem services.

Rationale

There are many reasons to restore ecosystems. Some include:

- Restoring natural capital such as drinkable water or wildlife populations.

- Mitigating climate change (e.g. through carbon sequestration)

- Helping threatened or endangered species

- Aesthetic reasons (Harris et al. 2006, Macdonald et al. 2002)

- Moral reasons: we have degraded, and in some cases destroyed, many ecosystems so it falls on us to 'fix' them.

There are considerable differences of opinion in how to set restoration goals and how to define their success. Some urge active restoration (e.g. eradicating invasive animals to allow the native ones to survive) and others who believe that protected areas should have the bare minimum of human interference. Ecosystem restoration has generated controversy, with skeptics who doubt that the benefits justify the economic investment or who point to failed restoration projects and question the feasibility of restoration altogether. It can be difficult to set restoration goals, in part because, as Anthony Bradshaw claims, "ecosystems are not static, but in a state of dynamic equilibrium.... [with restoration] we aim [for a] moving target."

Even though an ecosystem may not be returned to its original state, the functions of the ecosystem (especially ones that provide services to us) may be more valuable than its current configuration (Bradshaw 1987). One reason to consider ecosystem restoration is to mitigate climate change through activities such as afforestation. Afforestation involves replanting forests, which remove carbon dioxide from the air. Carbon dioxide is a leading cause of global warming (Speth, 2005) and capturing it would help alleviate climate change. Another example of a common driver of restoration projects in the United States is the legal framework of the Clean Water Act, which often requires mitigation for damage inflicted on aquatic systems by development or other activities.

Challenges in Restoration

Some view ecosystem restoration as impractical, in part because it sometimes fails. Hilderbrand et al. point out that many times uncertainty (about ecosystem functions, species relationships, and such) is not addressed, and that the time-scales set out for 'complete' restoration are unreasonably short. In other instances an ecosystem may be so degraded that abandonment (allowing an injured ecosystem to recover on its own) may be the wisest option (Holl, 2006). Local communities sometimes object to restorations that include the introduction of large predators or plants that require disturbance regimes such as regular fires (MacDonald et al. 2002). High economic costs can also be a perceived as a negative impact of the restoration process. Public opinion is very important in the feasibility of a restoration; if the public believes that the costs of restoration outweigh the benefits they will not support it (MacDonald et al. 2002). In these cases people might be ready to leave the ecosystem to recover on its own, which can sometimes occur relatively quickly (Holl, 2006).

Many failures have occurred in past restoration projects, many times because clear goals were not set out as the aim of the restoration. This may be because, as Peter Alpert says, "people may not [always] know how to manage natural systems effectively". Also many assumptions are made about myths of restoration such as the carbon copy, where a restoration plan, which worked in one area, is applied to another with the same results expected, but not realized (Hilderbrand et al. 2005).

Ecological Succession

Succession after disturbance: a boreal forest one year (left) and two years (right) after a wildfire.

Ecological succession is the process of change in the species structure of an ecological community over time. The time scale can be decades (for example, after a wildfire), or even millions of years after a mass extinction.

The community begins with relatively few pioneering plants and animals and develops through increasing complexity until it becomes stable or self-perpetuating as a climax community. The 'engine' of succession, the cause of ecosystem change, is the impact of established species upon their own environments. A consequence of living is the some-times subtle and sometimes overt alteration of one's own environment.

It is a phenomenon or process by which an ecological community undergoes more or less orderly and predictable changes following a disturbance or the initial colonization of a new habitat. Succession may be initiated either by formation of new, unoccupied habitat, such as from a lava flow or a severe landslide, or by some form of disturbance of a community, such as from a fire, severe windthrow, or logging. Succession that begins in new habitats, uninfluenced by pre-existing communities is called primary succession, whereas succession that follows disruption of a pre-existing community is called secondary succession.

Succession was among the first theories advanced in ecology. The study of succession remains at the core of ecological science. Ecological succession was first documented in the Indiana Dunes of Northwest Indiana which led to efforts to preserve the Indiana Dunes. Exhibits on ecological succession are displayed in the Hour Glass, a museum in Ogden Dunes.

History

Precursors of the idea of ecological succession go back to the beginning of the 19th century. The French naturalist Adolphe Dureau de la Malle was the first to make use of the word *succession* concerning the vegetation development after forest clear-cutting. In 1859 Henry David Thoreau wrote an address called "The Succession of Forest Trees" in which he described succession in an oak-pine forest. "It has long been known to observers that squirrels bury nuts in the ground, but I am not aware that any one has thus accounted for the regular succession of forests." The Austrian botanist Anton Kerner published a study about the succession of plants in the Danube river basin in 1863.

H. C. Cowles

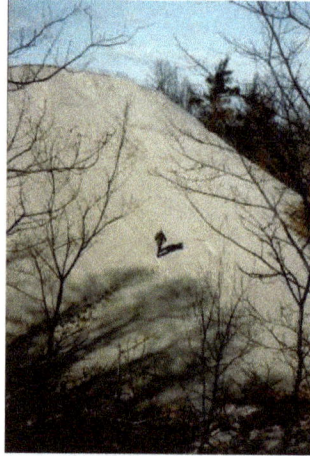

The Indiana Dunes on Lake Michigan, which stimulated Cowles' development of his theories of ecological succession

Henry Chandler Cowles, at the University of Chicago, developed a more formal concept of succession. Inspired by studies of Danish dunes by Eugen Warming, Cowles studied vegetation development on sand dunes on the shores of Lake Michigan (the Indiana Dunes). He recognized that vegetation on dunes of different ages might be interpreted as different stages of a general trend of vegetation development on dunes (an approach to the study of vegetation change later termed space-for-time substitution, or chronosequence studies). He first published this work as a paper in the *Botanical Gazette* in 1899 ("The ecological relations of the vegetation of the sand dunes of Lake Michigan"). In this classic publication and subsequent papers, he formulated the idea of primary succession and the notion of a sere—a repeatable sequence of community changes specific to particular environmental circumstances.

Gleason and Clements

From about 1900 to 1960, however, understanding of succession was dominated by the theories of Frederic Clements, a contemporary of Cowles, who held that seres were highly predictable and deterministic and converged on a climatically determined stable climax community regardless of starting conditions. Clements explicitly analogized the successional development of ecological communities with ontogenetic development of individual organisms, and his model is often referred to as the pseudo-organismic theory of community ecology. Clements and his followers developed a complex taxonomy of communities and successional pathways.

Henry Gleason offered a contrasting framework as early as the 1920s. The Gleasonian model was more complex and much less deterministic than the Clementsian. It differs most fundamentally from the Clementsian view in suggesting a much greater role of chance factors and in denying the existence of coherent, sharply bounded community types. Gleason argued that species distributions responded individualistically to environmental factors, and communities were best regarded as artifacts of the juxtaposition of species distributions. Gleason's ideas, first published in 1926, were largely ignored until the late 1950s.

Two quotes illustrate the contrasting views of Clements and Gleason. Clements wrote in 1916:

The developmental study of vegetation necessarily rests upon the assumption that the unit or climax formation is an organic entity. As an organism the formation arises, grows, matures, and dies. Furthermore, each climax formation is able to reproduce itself, repeating with essential fidelity the stages of its development.

— Frederic Clements

while Gleason, in his 1926 paper, said:

An association is not an organism, scarcely even a vegetational unit, but merely a coincidence.

— Henry Gleason

Gleason's ideas were, in fact, more consistent with Cowles' original thinking about succession. About Clements' distinction between primary succession and secondary succession, Cowles wrote (1911):

This classification seems not to be of fundamental value, since it separates such closely related phenomena as those of erosion and deposition, and it places together such unlike things as human agencies and the subsidence of land.

— Henry Cowles

Modern Era

A more rigorous, data-driven testing of successional models and community theory generally began with the work of Robert Whittaker and John Curtis in the 1950s and 1960s. Succession theory has since become less monolithic and more complex. J. Connell and R. Slatyer attempted a codification of successional processes by mechanism. Among British and North American ecologists, the notion of a stable climax vegetation has been largely abandoned, and successional processes have come to be seen as much less deterministic, with important roles for historical contingency and for alternate pathways in the actual development of communities. Debates continue as to the general predictability of successional dynamics and the relative importance of equilibrial vs. non-equilibrial processes. Former Harvard professor F. A. Bazzaz introduced the notion of *scale* into the discussion, as he considered that at local or small area scale the processes are stochastic and patchy, but taking bigger regional areas into consideration, certain tendencies can not be denied.

Factors

The trajectory of successional change can be influenced by site conditions, by the character of the events initiating succession (perturbations), by the interactions of the species present, and by more stochastic factors such as availability of colonists or seeds or weather conditions at the time of disturbance. Some of these factors contribute to predictability of succession dynamics; others add more probabilistic elements. Two important perturbation factors today are human actions and climatic change.

In general, communities in early succession will be dominated by fast-growing, well-dispersed species (opportunist, fugitive, or r-selected life-histories). As succession proceeds, these species will tend to be replaced by more competitive (k-selected) species.

Trends in ecosystem and community properties in succession have been suggested, but few appear to be general. For example, species diversity almost necessarily increases during early succession as new species arrive, but may decline in later succession as competition eliminates opportunistic species and leads to dominance by locally superior competitors. Net Primary Productivity, biomass, and trophic properties all show variable patterns over succession, depending on the particular system and site.

Ecological succession was formerly seen as having a stable end-stage called the climax, sometimes referred to as the 'potential vegetation' of a site, and shaped primarily by the local climate. This idea has been largely abandoned by modern ecologists in favor of nonequilibrium ideas of ecosystems dynamics. Most natural ecosystems experience disturbance at a rate that makes a "climax" community unattainable. Climate change often occurs at a rate and frequency sufficient to prevent arrival at a climax state. Additions to available species pools through range expansions and introductions can also continually reshape communities.

The development of some ecosystem attributes, such as soil properties and nutrient cycles, are both influenced by community properties, and, in turn, influence further successional development. This feed-back process may occur only over centuries or millennia. Coupled with the stochastic nature of disturbance events and other long-term (e.g., climatic) changes, such dynamics make it doubtful whether the 'climax' concept ever applies or is particularly useful in considering actual vegetation.

Types

Primary, Secondary and Cyclic Succession

An example of Secondary Succession by stages:

1. A stable deciduous forest community

2. A disturbance, such as a wild fire, destroys the forest

3. The fire burns the forest to the ground

4. The fire leaves behind empty, but not destroyed, soil

5. Grasses and other herbaceous plants grow back first

6. Small bushes and trees begin to colonize the area

7. Fast growing evergreen trees develop to their fullest, while shade-tolerant trees develop in the under-story

8. The short-lived and shade intolerant evergreen trees die as the larger deciduous trees overtop them. The ecosystem is now back to a similar state to where it began.

Successional dynamics beginning with colonization of an area that has not been previously occupied by an ecological community, such as newly exposed rock or sand surfaces, lava flows, newly exposed glacial tills, etc., are referred to as primary succession. The stages of primary succession include pioneer plants (lichens and mosses), grassy stage, smaller shrubs, and trees. Animals begin to return when there is food there for them to eat. When it is a fully functioning ecosystem, it has reached the climax community stage. For example, parts of Acadia National Park in Maine went through primary succession.

Secondary succession: trees are colonizing uncultivated fields and meadows.

Successional dynamics following severe disturbance or removal of a pre-existing community are called secondary succession. Dynamics in secondary succession are strongly influenced by pre-disturbance conditions, including soil development, seed banks, remaining organic matter, and residual living organisms. Because of residual fertility and pre-existing organisms, community change in early stages of secondary succession can be relatively rapid. In a fragmented old field habitat created in eastern Kansas, woody plants "colonized more rapidly (per unit area) on large and nearby patches."

Secondary succession is much more commonly observed and studied than primary succession. Particularly common types of secondary succession include responses to natural disturbances such as fire, flood, and severe winds, and to human-caused disturbances such as logging and agriculture. As an example, secondary succession has been occurring in Shenandoah National Park following the 1995 flood of the Mormon River, which destroyed plant and animal life. Today, plant and animal species are beginning to return.

Seasonal and Cyclic Dynamics

Unlike secondary succession, these types of vegetation change are not dependent on disturbance but are periodic changes arising from fluctuating species interactions or recurring events. These models modify the climax concept towards one of dynamic states.

Causes of Plant Succession

Autogenic succession can be brought by changes in the soil caused by the organisms there. These changes include accumulation of organic matter in litter or humic layer, alteration of soil nutrients, change in pH of soil by plants growing there. The structure of the plants themselves can also alter the community. For example, when larger species like trees mature, they produce shade on to the developing forest floor that tends to exclude light-requiring species. Shade-tolerant species will invade the area.

Allogenic succession is caused by external environmental influences and not by the vegetation. For example, soil changes due to erosion, leaching or the deposition of silt and clays can alter the nutrient content and water relationships in the ecosystems. Animals also play an important role in allogenic changes as they are pollinators, seed dispersers and herbivores. They can also increase nutrient content of the soil in certain areas, or shift soil about (as termites, ants, and moles do) creating patches in the habitat. This may create regeneration sites that favor certain species.

Climatic factors may be very important, but on a much longer time-scale than any other. Changes in temperature and rainfall patterns will promote changes in communities. As the climate warmed at the end of each ice age, great successional changes took place. The tundra vegetation and bare glacial till deposits underwent succession to mixed deciduous forest. The greenhouse effect resulting in increase in temperature is likely to bring profound Allogenic changes in the next century. Geological and climatic catastrophes such as volcanic eruptions, earthquakes, avalanches, meteors, floods, fires, and high wind also bring allogenic changes.

Mechanisms

In 1916, Frederic Clements published a descriptive theory of succession and advanced it as a general ecological concept. His theory of succession had a powerful influence on ecological thought. Clements' concept is usually termed classical ecological theory. According to Clements, succession is a process involving several phases:

1. Nudation: Succession begins with the development of a bare site, called Nudation (disturbance).

2. Migration: It refers to arrival of propagules.

3. Ecesis: It involves establishment and initial growth of vegetation.

4. Competition: As vegetation becomes well established, grow, and spread, various species begin to compete for space, light and nutrients.

5. Reaction: During this phase autogenic changes such as the buildup of humus affect the habitat, and one plant community replaces another.

6. Stabilization: A supposedly stable climax community forms.

Seral Communities

A hydrosere community

A seral community is an intermediate stage found in an ecosystem advancing towards its climax community. In many cases more than one seral stage evolves until climax conditions are attained. A *prisere* is a collection of seres making up the development of an area from non-vegetated surfaces to a climax community. Depending on the substratum and climate, different seres are found.

Changes in Animal Life

Succession theory was developed primarily by botanists. The study of succession applied to whole ecosystems initiated in the writings of Ramon Margalef, while Eugene Odum's publication of *The Strategy of Ecosystem Development* is considered its formal starting point.

Animal life also exhibit changes with changing communities. In lichen stage the fauna is sparse. It comprises few mites, ants and spiders living in the cracks and crevices. The fauna undergoes a qualitative increase during herb grass stage. The animals found during this stage include nematodes, insects larvae, ants, spiders, mites, etc. The animal population increases and diversifies with the development of forest climax community. The fauna consists of invertebrates like slugs, snails, worms, millipedes, centipedes, ants, bugs; and vertebrates such as squirrels, foxes, mice, moles, snakes, various birds, salamanders and frogs.

Microsuccession

Succession of micro-organisms including fungi and bacteria occurring within a microhabitat is known as microsuccession or serule. This type of succession occurs in recently disturbed communities or newly available habitat, for example in recently dead trees, animal droppings, exposed glacial till, etc. Microbial communities may also change due to products secreted by the bacteria present. Changes of pH in a habitat could provide ideal conditions for a new species to inhabit the area. In some cases the new species may outcompete the present ones for nutrients leading to the primary species demise. Changes can also occur by microbial succession with variations in water availability and temperature. Theories of macroecology have only recently been applied to microbiology and so much remains to be understood about this growing field. A recent study of microbial succession evaluated the balances between stochastic and deterministic processes in the bacterial colonization of a salt marsh chronosequence. The results of this study show that, much like in macro succession, early colonization (primary succession) is mostly influenced by stochasticity while secondary succession of these bacterial communities was more strongly influenced by deterministic factors.

Climax Concept

According to classical ecological theory, succession stops when the sere has arrived at an equilibrium or steady state with the physical and biotic environment. Barring major disturbances, it will persist indefinitely. This end point of succession is called climax.

Climax Community

The final or stable community in a sere is the *climax community* or *climatic vegetation*. It is self-perpetuating and in equilibrium with the physical habitat. There is no net annual accumulation of organic matter in a climax community. The annual production and use of energy is balanced in such a community.

Characteristics

- The vegetation is tolerant of environmental conditions.

- It has a wide diversity of species, a well-drained spatial structure, and complex food chains.

- The climax ecosystem is balanced. There is equilibrium between gross primary production and total respiration, between energy used from sunlight and energy released by decomposition, between uptake of nutrients from the soil and the return of nutrient by litter fall to the soil.

- Individuals in the climax stage are replaced by others of the same kind. Thus the species composition maintains equilibrium.

- It is an index of the climate of the area. The life or growth forms indicate the climatic type.

Types of Climax

Climatic Climax

If there is only a single climax and the development of climax community is controlled by the climate of the region, it is termed as climatic climax. For example, development of Maple-beech climax community over moist soil. Climatic climax is theoretical and develops where physical conditions of the substrate are not so extreme as to modify the effects of the prevailing regional climate.

Edaphic Climax

When there are more than one climax communities in the region, modified by local conditions of the substrate such as soil moisture, soil nutrients, topography, slope exposure, fire, and animal activity, it is called *edaphic climax*. Succession ends in an edaphic climax where topography, soil, water, fire, or other disturbances are such that a climatic climax cannot develop.

Catastrophic Climax

Climax vegetation vulnerable to a catastrophic event such as a wildfire. For example, in California, chaparral vegetation is the final vegetation. The wildfire removes the mature vegetation and decomposers. A rapid development of herbaceous vegetation follows until the shrub dominance is re-established. This is known as catastrophic climax.

Disclimax

When a stable community, which is not the climatic or edaphic climax for the given site, is maintained by man or his domestic animals, it is designated as Dis-climax (disturbance climax) or anthropogenic subclimax (man-generated). For example, overgrazing by stock may produce a desert community of bushes and cacti where the local climate actually would allow grassland to maintain itself.

Subclimax

The prolonged stage in succession just preceding the climatic climax is *subclimax*.

Preclimax and Postclimax

In certain areas different climax communities develop under similar climatic conditions. If the community has life forms lower than those in the expected climatic climax, it is called *preclimax*; a community that has life forms higher than those in the expected climatic climax is *postclimax*. Preclimax strips develop in less moist and hotter areas, whereas Postclimax strands develop in more moist and cooler areas than that of surrounding climate.

Theories

There are three schools of interpretations explaining the climax concept:

- Monoclimax or Climatic Climax Theory was advanced by Clements (1916) and recognizes only one climax whose characteristics are determined solely by climate (climatic climax). The processes of succession and modification of environment overcome the effects of differences in topography, parent material of the soil, and other factors. The whole area would be covered with uniform plant community. Communities other than the climax are related to it, and are recognized as subclimax, postclimax and disclimax.

- Polyclimax Theory was advanced by Tansley (1935). It proposes that the climax vegetation of a region consists of more than one vegetation climaxes controlled by soil moisture, soil nutrients, topography, slope exposure, fire, and animal activity.

- Climax Pattern Theory was proposed by Whittaker (1953). The climax pattern theory recognizes a variety of climaxes governed by responses of species populations to biotic and abiotic conditions. According to this theory the total environment of the ecosystem determines the composition, species structure, and balance of a climax community. The environment includes the species responses to moisture, temperature, and nutrients, their biotic relationships, availability of flora and fauna to colonize the area, chance dispersal of seeds and animals, soils, climate, and disturbance such as fire and wind. The nature of climax vegetation will change as the environment changes. The climax community represents a pattern of populations that corresponds to and changes with the pattern of environment. The central and most widespread community is the climatic climax.

More recently another possible idea has been put forward called the theory of alternative stable states which suggests that there is not one end point but many which transition between each other over ecological time.

Forest Succession

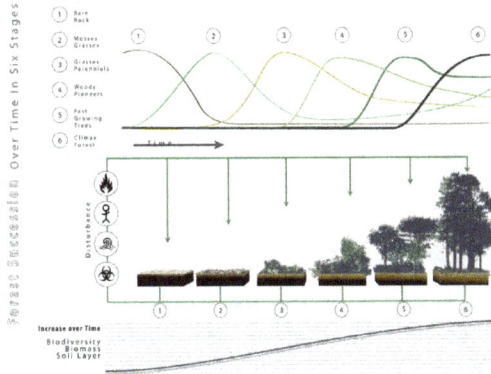

The forests, being an ecological system, are subject to the species succession process. There are "opportunistic" or "pioneer" species that produce great quantities of seed that are disseminated by the wind, and therefore can colonize big empty extensions. They are capable of germinating and growing in direct sunlight. Once they have produced a *closed canopy*, the lack of direct sun radiation at soil makes it difficult for their own seedlings to develop. It is then the opportunity for shade-tolerant species to become established under the protection of the pioneers. When the pioneers die, the shade-tolerant species replace them. These species are capable of growing beneath the canopy, and therefore, in the absence of catastrophes, will stay. For this reason it is then said the stand has reached its climax. When a catastrophe occurs, the opportunity for the pioneers opens up again, provided they are present or within a reasonable range.

An example of pioneer species, in forests of northeastern North America are *Betula papyrifera* (White birch) and *Prunus serotina* (Black cherry), that are particularly well-adapted to exploit large gaps in forest canopies, but are intolerant of shade and are eventually replaced by other shade-tolerant species in the absence of disturbances that create such gaps.

Things in nature are not black and white, and there are intermediate stages. It is therefore normal that between the two extremes of light and shade there is a gradient, and there are species that may act as pioneer or tolerant, depending on the circumstances. It is of paramount importance to know the tolerance of species in order to practice an effective silviculture..

References

- Eric Laferrière; Peter J. Stoett (2 September 2003). International Relations Theory and Ecological Thought: Towards a Synthesis. Routledge. pp. 25–. ISBN 978-1-134-71068-3.

- O'Neill, D. L.; Deangelis, D. L.; Waide, J. B.; Allen, T. F. H. (1986). A Hierarchical Concept of Ecosystems. Princeton University Press. p. 253. ISBN 0-691-08436-X.

- Noss, R. F.; Carpenter, A. Y. (1994). Saving Nature's Legacy: Protecting and Restoring Biodiversity. Island Press. p. 443. ISBN 978-1-55963-248-5.

- Hammond, H. (2009). Maintaining Whole Systems on the Earth's Crown: Ecosystem-based Conservation Planning for the Boreal Forest. Slocan Park, BC: Silva Forest Foundation. p. 380. ISBN 978-0-9734779-0-0.

- Vandermeer, J. H.; Goldberg, D. E. (2003). Population Ecology: First Principles. Woodstock, Oxfordshire: Princeton University Press. ISBN 0-691-11440-4.

- Levins, R. (1970). "Extinction". In Gerstenhaber, M. Some Mathematical Questions in Biology. pp. 77–107. ISBN 978-0-8218-1152-8.

- Hanski, I.; Gaggiotti, O. E., eds. (2004). Ecology, Genetics and Evolution of Metapopulations. Burlington, MA: Elsevier Academic Press. ISBN 0-12-323448-4.

- MacKenzie; D.I. (2006). Occupancy Estimation and Modeling: Inferring Patterns and Dynamics of Species Occurrence. London, UK: Elsevier Academic Press. p. 324. ISBN 978-0-12-088766-8.

- Allee, W. C.; Park, O.; Emerson, A. E.; Park, T.; Schmidt, K. P. (1949). Principles of Animal Ecology. W. B. Sunders, Co. p. 837. ISBN 0-7216-1120-6.

- Francisco J Varela; Evan Thompson; Eleanor Rosch (1993). The Embodied Mind: Cognitive Science and Human Experience (Paperback ed.). MIT Press. p. 174. ISBN 9780262261234.

- Campbell, Neil A.; Williamson, Brad; Heyden, Robin J. (2006). Biology: Exploring Life. Boston, Massachusetts: Pearson Prentice Hall. ISBN 0-13-250882-6.

6

Diverse Applications of Agroecology

Agroecology has a number of diverse applications. Some of these are organic farming, organic food, conservation agriculture, shifting cultivation and intercropping. Organic farming depends on fertilizers such as manure, compost and green manure. The food produced through the methods of organic farming is known ad organic food. This section serves as a source to understand the major categories related to agroecology.

Organic Farming

Organic farming is an alternative agricultural system which originated early in the 20th century in reaction to rapidly changing farming practices. Organic agriculture continues to be developed by various organic agriculture organizations today. It relies on fertilizers of organic origin such as compost, manure, green manure, and bone meal and places emphasis on techniques such as crop rotation and companion planting. Biological pest control, mixed cropping and the fostering of insect predators are encouraged. In general, organic standards are designed to allow the use of naturally occurring substances while prohibiting or strictly limiting synthetic substances. For instance, naturally occurring pesticides such as pyrethrin and rotenone are permitted, while synthetic fertilizers and pesticides are generally prohibited. Synthetic substances that are allowed include, for example, copper sulfate, elemental sulfur and Ivermectin. Genetically modified organisms, nanomaterials, human sewage sludge, plant growth regulators, hormones, and antibiotic use in livestock husbandry are prohibited. Reasons for advocation of organic farming include real or perceived advantages in sustainability, openness, self-sufficiency, autonomy/independence, health, food security, and food safety, although the match between perception and reality is continually challenged.

World map of organic agriculture (hectares)

Vegetables from ecological farming.

Organic agricultural methods are internationally regulated and legally enforced by many nations, based in large part on the standards set by the International Federation of Organic Agriculture Movements (IFOAM), an international umbrella organization for organic farming organizations established in 1972. Organic agriculture can be defined as:

an integrated farming system that strives for sustainability, the enhancement of soil fertility and biological diversity whilst, with rare exceptions, prohibiting synthetic pesticides, antibiotics, synthetic fertilizers, genetically modified organisms, and growth hormones.

Since 1990 the market for organic food and other products has grown rapidly, reaching $63 billion worldwide in 2012. This demand has driven a similar increase in organically managed farmland that grew from 2001 to 2011 at a compounding rate of 8.9% per annum. As of 2011, approximately 37,000,000 hectares (91,000,000 acres) worldwide were farmed organically, representing approximately 0.9 percent of total world farmland.

History

Agriculture was practiced for thousands of years without the use of artificial chemicals. Artificial fertilizers were first created during the mid-19th century. These early fertilizers were cheap, powerful, and easy to transport in bulk. Similar advances occurred in chemical pesticides in the 1940s, leading to the decade being referred to as the 'pesticide era'. These new agricultural techniques, while beneficial in the short term, had serious longer term side effects such as soil compaction, erosion, and declines in overall soil fertility, along with health concerns about toxic chemicals entering the food supply. In the late 1800s and early 1900s, soil biology scientists began to seek ways to remedy these side effects while still maintaining higher production.

Biodynamic agriculture was the first modern system of agriculture to focus exclusively on organic methods. Its development began in 1924 with a series of eight lectures on agriculture given by Rudolf Steiner. These lectures, the first known presentation of what later came to be known as organic agriculture, were held in response to a request by farmers who noticed degraded soil conditions and a deterioration in the health and quality of crops and livestock resulting from the use of chemical fertilizers. The one hundred eleven attendees, less than half of whom were farmers, came from six countries, primarily Germany and Poland. The lectures were published in November 1924; the first English translation appeared in 1928 as *The Agriculture Course*.

In 1921, Albert Howard and his wife Gabrielle Howard, accomplished botanists, founded an Institute of Plant Industry to improve traditional farming methods in India. Among other things, they brought improved implements and improved animal husbandry methods from their scientific training; then by incorporating aspects of the local traditional methods, developed protocalls for the rotation of crops, erosion prevention techniques, and the systematic use of composts and manures. Stimulated by these experiences of traditional farming, when Albert Howard returned to Britain in the early 1930s he began to promulgate a system of natural agriculture.

In July 1939, Ehrenfried Pfeiffer, the author of the standard work on biodynamic agriculture (*Bio-Dynamic Farming and Gardening*), came to the UK at the invitation of Walter James, 4th Baron Northbourne as a presenter at the Betteshanger Summer School and Conference on Biodynamic Farming at Northbourne's farm in Kent. One of the chief purposes of the conference was to bring together the proponents of various approaches to organic agriculture in order that they might cooperate within a larger movement. Howard attended the conference, where he met Pfeiffer. In the following year, Northbourne published his manifesto of organic farming, *Look to the Land*, in which he coined the term "organic farming." The Betteshanger conference has been described as the 'missing link' between biodynamic agriculture and other forms of organic farming.

In 1940 Howard published his *An Agricultural Testament*. In this book he adopted Northbourne's terminology of "organic farming." Howard's work spread widely, and he became known as the "father of organic farming" for his work in applying scientific knowledge and principles to various traditional and natural methods. In the United States J.I. Rodale, who was keenly interested both in Howard's ideas and in biodynamics, founded in the 1940s both a working organic farm for trials and experimentation, The Rodale Institute, and the Rodale Press to teach and advocate organic methods to the wider public. These became important influences on the spread of organic agriculture. Further work was done by Lady Eve Balfour in the United Kingdom, and many others across the world.

Increasing environmental awareness in the general population in modern times has transformed the originally supply-driven organic movement to a demand-driven one. Premium prices and some government subsidies attracted farmers. In the developing world, many producers farm according to traditional methods that are comparable to organic farming, but not certified, and that may not include the latest scientific advancements in organic agriculture. In other cases, farmers in the developing world have converted to modern organic methods for economic reasons.

Terminology

Biodynamic agriculturists, who based their work on Steiner's spiritually-oriented anthroposophy, used the term "organic" to indicate that a farm should be viewed as a living organism, in the sense of the following quotation:

"An organic farm, properly speaking, is not one that uses certain methods and substances and avoids others; it is a farm whose structure is formed in imitation of the structure of a natural system that has the integrity, the independence and the benign dependence of an organism"

— Wendell Berry, "The Gift of Good Land"

The use of "organic" popularized by Howard and Rodale, on the other hand, refers more narrowly to the use of organic matter derived from plant compost and animal manures to improve the humus content of soils, grounded in the work of early soil scientists who developed what was then called "humus farming." Since the early 1940s the two camps have tended to merge.

Methods

Organic cultivation of mixed vegetables in Capay, California. Note the hedgerow in the background.

"Organic agriculture is a production system that sustains the health of soils, ecosystems and people. It relies on ecological processes, biodiversity and cycles adapted to local conditions, rather than the use of inputs with adverse effects. Organic agriculture combines tradition, innovation and science to benefit the shared environment and promote fair relationships and a good quality of life for all involved..."

— International Federation of Organic Agriculture Movements

Organic farming methods combine scientific knowledge of ecology and modern technology with traditional farming practices based on naturally occurring biological processes. Organic farming methods are studied in the field of agroecology. While conventional agriculture uses synthetic pesticides and water-soluble synthetically purified fertilizers, organic farmers are restricted by regulations to using natural pesticides and fertilizers. An example of a natural pesticide is pyrethrin, which is found naturally in the Chrysanthemum flower. The principal methods of organic farming include crop rotation, green manures and compost, biological pest control, and mechanical cultivation. These measures use the natural environment to enhance agricultural productivity: legumes are planted to fix nitrogen into the soil, natural insect predators are encouraged, crops are rotated to confuse pests and renew soil, and natural materials such as potassium bicarbonate and mulches are used to control disease and weeds. Genetically modified seeds and animals are excluded.

While organic is fundamentally different from conventional because of the use of carbon based fertilizers compared with highly soluble synthetic based fertilizers and biological pest control instead of synthetic pesticides, organic farming and large-scale conventional farming are not entirely mutually exclusive. Many of the methods developed for organic agriculture have been borrowed by more conventional agriculture. For example, Integrated Pest Management is a multifaceted strategy that uses various organic methods of pest control whenever possible, but in conventional farming could include synthetic pesticides only as a last resort.

Crop Diversity

Organic farming encourages Crop diversity. The science of agroecology has revealed the benefits of polyculture (multiple crops in the same space), which is often employed in organic farming. Planting a variety of vegetable crops supports a wider range of beneficial insects, soil microorganisms, and other factors that add up to overall farm health. Crop diversity helps environments thrive and protects species from going extinct.

Soil Management

Organic farming relies heavily on the natural breakdown of organic matter, using techniques like green manure and composting, to replace nutrients taken from the soil by previous crops. This biological process, driven by microorganisms such as mycorrhiza, allows the natural production of nutrients in the soil throughout the growing season, and has been referred to as *feeding the soil to feed the plant*. Organic farming uses a variety of methods to improve soil fertility, including crop rotation, cover cropping, reduced tillage, and application of compost. By reducing tillage, soil is not inverted and exposed to air; less carbon is lost to the atmosphere resulting in more soil organic carbon. This has an added benefit of carbon sequestration, which can reduce green house gases and help reverse climate change.

Plants need nitrogen, phosphorus, and potassium, as well as micronutrients and symbiotic relationships with fungi and other organisms to flourish, but getting enough nitrogen, and particularly synchronization so that plants get enough nitrogen at the right time (when plants need it most), is a challenge for organic farmers. Crop rotation and green manure ("cover crops") help to provide nitrogen through legumes (more precisely, the *Fabaceae* family), which fix nitrogen from the atmosphere through symbiosis with rhizobial bacteria. Intercropping, which is sometimes used for insect and disease control, can also increase soil nutrients, but the competition between the legume and the crop can be problematic and wider spacing between crop rows is required. Crop residues can be ploughed back into the soil, and different plants leave different amounts of nitrogen, potentially aiding synchronization. Organic farmers also use animal manure, certain processed fertilizers such as seed meal and various mineral powders such as rock phosphate and green sand, a naturally occurring form of potash that provides potassium. Together these methods help to control erosion. In some cases pH may need to be amended. Natural pH amendments include lime and sulfur, but in the U.S. some compounds such as iron sulfate, aluminum sulfate, magnesium sulfate, and soluble boron products are allowed in organic farming.

Mixed farms with both livestock and crops can operate as ley farms, whereby the land gathers fertility through growing nitrogen-fixing forage grasses such as white clover or alfalfa and grows cash crops or cereals when fertility is established. Farms without livestock ("stockless") may find it more difficult to maintain soil fertility, and may rely more on external inputs such as imported manure as well as grain legumes and green manures, although grain legumes may fix limited nitrogen because they are harvested. Horticultural farms that grow fruits and vegetables in protected conditions often relay even more on external inputs.

Biological research into soil and soil organisms has proven beneficial to organic farming. Varieties of bacteria and fungi break down chemicals, plant matter and animal waste into productive soil nutrients. In turn, they produce benefits of healthier yields and more productive soil for future crops. Fields with less or no manure display significantly lower yields, due to decreased soil microbe community. Increased manure improves biological activity, providing a healthier, more arable soil system and higher yields.

Weed Management

Organic weed management promotes weed suppression, rather than weed elimination, by enhancing crop competition and phytotoxic effects on weeds. Organic farmers integrate cultural, biological, mechanical, physical and chemical tactics to manage weeds without synthetic herbicides.

Organic standards require rotation of annual crops, meaning that a single crop cannot be grown in the same location without a different, intervening crop. Organic crop rotations frequently include weed-suppressive cover crops and crops with dissimilar life cycles to discourage weeds associated with a particular crop. Research is ongoing to develop organic methods to promote the growth of natural microorganisms that suppress the growth or germination of common weeds.

Other cultural practices used to enhance crop competitiveness and reduce weed pressure include selection of competitive crop varieties, high-density planting, tight row spacing, and late planting into warm soil to encourage rapid crop germination.

Mechanical and physical weed control practices used on organic farms can be broadly grouped as:

- Tillage - Turning the soil between crops to incorporate crop residues and soil amendments; remove existing weed growth and prepare a seedbed for planting; turning soil after seeding to kill weeds, including cultivation of row crops;

- Mowing and cutting - Removing top growth of weeds;

- Flame weeding and thermal weeding - Using heat to kill weeds; and

- Mulching - Blocking weed emergence with organic materials, plastic films, or landscape fabric.

Some critics, citing work published in 1997 by David Pimentel of Cornell University, which described an epidemic of soil erosion worldwide, have raised concerned that tillage contribute to the erosion epidemic. The FAO and other organizations have advocated a 'no-till' approach to both conventional and organic farming, and point out in particular that crop rotation techniques used in organic farming are excellent no-till approaches. A study published in 2005 by Pimentel and colleagues confirmed that 'Crop rotations and cover cropping (green manure) typical of organic agriculture reduce soil erosion, pest problems, and pesticide use.' Some naturally sourced chemicals are allowed for herbicidal use. These include certain formulations of acetic acid (concentrated vinegar), corn gluten meal, and essential oils. A few selective bioherbicides based on fungal pathogens have also been developed. At this time, however, organic herbicides and bioherbicides play a minor role in the organic weed control toolbox.

Weeds can be controlled by grazing. For example, geese have been used successfully to weed a range of organic crops including cotton, strawberries, tobacco, and corn, reviving the practice of keeping cotton patch geese, common in the southern U.S. before the 1950s. Similarly, some rice farmers introduce ducks and fish to wet paddy fields to eat both weeds and insects.

Controlling other Organisms

Organisms aside from weeds that cause problems on organic farms include arthropods (e.g., insects, mites), nematodes, fungi and bacteria. Organic practices include, but are not limited to:

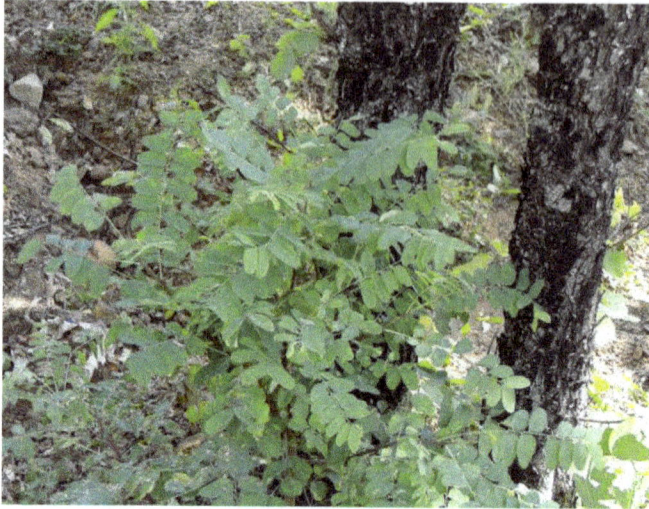

Chloroxylon is used for Pest Management in Organic Rice Cultivation in Chhattisgarh, India

- encouraging predatory beneficial insects to control pests by serving them nursery plants and/or an alternative habitat, usually in a form of a shelterbelt, hedgerow, or beetle bank;

- encouraging beneficial microorganisms;

- rotating crops to different locations from year to year to interrupt pest reproduction cycles;

- planting companion crops and pest-repelling plants that discourage or divert pests;

- using row covers to protect crops during pest migration periods;

- using biologic pesticides and herbicides

- using stale seed beds to germinate and destroy weeds before planting

- using sanitation to remove pest habitat;

- Using insect traps to monitor and control insect populations.

- Using physical barriers, such as row covers

Examples of predatory beneficial insects include minute pirate bugs, big-eyed bugs, and to a lesser extent ladybugs (which tend to fly away), all of which eat a wide range of pests. Lacewings are also effective, but tend to fly away. Praying mantis tend to move more slowly and eat less heavily. Parasitoid wasps tend to be effective for their selected prey, but like all small insects can be less effective outdoors because the wind controls their movement. Predatory mites are effective for controlling other mites.

Naturally derived insecticides allowed for use on organic farms use include *Bacillus thuringiensis* (a bacterial toxin), pyrethrum (a chrysanthemum extract), spinosad (a bacterial metabolite), neem (a tree extract) and rotenone (a legume root extract). Fewer than 10% of organic farmers use these pesticides regularly; one survey found that only 5.3% of vegetable growers in California use rotenone while 1.7% use pyrethrum. These pesticides are not always more safe or environmentally friendly than synthetic pesticides and can cause harm. The main criterion for organic pesticides is that they are naturally derived, and some naturally derived substances have been controversial. Controversial natural pesticides include rotenone, copper, nicotine sulfate, and pyrethrums Rotenone and pyrethrum are particularly controversial because they work by attacking the nervous system, like most conventional insecticides. Rotenone is extremely toxic to fish and can induce symptoms resembling Parkinson's disease in mammals. Although pyrethrum (natural pyrethrins) is more effective against insects when used with piperonyl butoxide (which retards degradation of the pyrethrins), organic standards generally do not permit use of the latter substance.

Naturally derived fungicides allowed for use on organic farms include the bacteria *Bacillus subtilis* and *Bacillus pumilus*; and the fungus *Trichoderma harzianum*. These are mainly effective for diseases affecting roots. Compost tea contains a mix of beneficial microbes, which may attack or out-compete certain plant pathogens, but variability among formulations and preparation methods may contribute to inconsistent results or even dangerous growth of toxic microbes in compost teas.

Some naturally derived pesticides are not allowed for use on organic farms. These include nicotine sulfate, arsenic, and strychnine.

Synthetic pesticides allowed for use on organic farms include insecticidal soaps and horticultural oils for insect management; and Bordeaux mixture, copper hydroxide and sodium bicarbonate for managing fungi. Copper sulfate and Bordeaux mixture (copper sulfate plus lime), approved for organic use in various jurisdictions, can be more environmentally problematic than some synthetic fungicides dissallowed in organic farming Similar concerns apply to copper hydroxide. Repeated application of copper sulfate or copper hydroxide as a fungicide may eventually result in copper accumulation to toxic levels in soil, and admonitions to avoid excessive accumulations of copper in soil appear in various organic standards and elsewhere. Environmental concerns for several kinds of biota arise at average rates of use of such substances for some crops. In the European Union, where replacement of copper-based fungicides in organic agriculture is a policy priority, research is seeking alternatives for organic production.

Livestock

Raising livestock and poultry, for meat, dairy and eggs, is another traditional farming activity that complements growing. Organic farms attempt to provide animals with natural living conditions and feed. Organic certification verifies that livestock are raised according to the USDA organic regulations throughout their lives. These regulations include the requirement that all animal feed must be certified organic.

For livestock like these healthy cows vaccines play an important part in animal health since antibiotic therapy is prohibited in organic farming

Organic livestock may be, and must be, treated with medicine when they are sick, but drugs cannot be used to promote growth, their feed must be organic, and they must be pastured.

Also, horses and cattle were once a basic farm feature that provided labor, for hauling and plowing, fertility, through recycling of manure, and fuel, in the form of food for farmers and other animals. While today, small growing operations often do not include livestock, domesticated animals are a desirable part of the organic farming equation, especially for true sustainability, the ability of a farm to function as a self-renewing unit.

Genetic Modification

A key characteristic of organic farming is the rejection of genetically engineered plants and animals. On 19 October 1998, participants at IFOAM's 12th Scientific Conference issued the Mar del Plata Declaration, where more than 600 delegates from over 60 countries voted unanimously to exclude the use of genetically modified organisms in food production and agriculture.

Although opposition to the use of any transgenic technologies in organic farming is strong, agricultural researchers Luis Herrera-Estrella and Ariel Alvarez-Morales continue to advocate integration of transgenic technologies into organic farming as the optimal means to sustainable agriculture, particularly in the developing world, as does author and scientist Pamela Ronald, who views this kind of biotechnology as being consistent with organic principles.

Although GMOs are excluded from organic farming, there is concern that the pollen from genetically modified crops is increasingly penetrating organic and heirloom seed stocks, making it difficult, if not impossible, to keep these genomes from entering the organic food supply. Differing regulations among countries limits the availability of GMOs to certain countries, as described in the article on regulation of the release of genetic modified organisms.

Tools

Organic farmers use a number of traditional farm tools to do farming. Due to the goals of sustainability in organic farming, organic farmers try to minimize their reliance on fossil fuels. In the developing world on small organic farms tools are normally constrained to hand tools and diesel powered water pumps. A recent study evaluated the use of open-source 3-D printers (called RepRaps using a bioplastic polylactic acid (PLA) on organic farms. PLA is a strong biodegradable and recyclable thermoplastic appropriate for a range of representative products in five categories of prints: handtools, food processing, animal management, water management and hydroponics. Such open source hardware is attractive to all types of small farmers as it provides control for farmers over their own equipment; this is exemplified by Open Source Ecology, Farm Hack and FarmBot.

Standards

Standards regulate production methods and in some cases final output for organic agriculture. Standards may be voluntary or legislated. As early as the 1970s private associations certified organic producers. In the 1980s, governments began to produce organic production guidelines. In the 1990s, a trend toward legislated standards began, most notably with the 1991 EU-Eco-regulation developed for European Union, which set standards for 12 countries, and a 1993 UK program. The EU's program was followed by a Japanese program in 2001, and in 2002 the U.S. created the National Organic Program (NOP). As of 2007 over 60 countries regulate organic farming (IFOAM 2007:11). In 2005 IFOAM created the Principles of Organic Agriculture, an international guideline for certification criteria. Typically the agencies accredit certification groups rather than individual farms.

Organic production materials used in and foods are tested independently by the Organic Materials Review Institute.

Composting

Using manure as a fertiliser risks contaminating food with animal gut bacteria, including pathogenic strains of E. coli that have caused fatal poisoning from eating organic food. To combat this risk, USDA organic standards require that manure must be sterilized through high temperature thermophilic composting. If raw animal manure is used, 120 days must pass before the crop is harvested if the final product comes into direct contact with the soil. For products that don't directly contact soil, 90 days must pass prior to harvest.

Economics

The economics of organic farming, a subfield of agricultural economics, encompasses the entire process and effects of organic farming in terms of human society, including social costs, opportunity costs, unintended consequences, information asymmetries, and economies of scale. Although the scope of economics is broad, agricultural economics tends to focus on maximizing yields and efficiency at the farm level. Economics takes an anthropocentric approach to the value of the natural world: biodiversity, for example, is considered beneficial only to the extent that it is valued by people and increases profits. Some entities such as the European Union subsidize organic farming, in large part because these countries want to account for the externalities of reduced water use, reduced water contamination, reduced soil erosion, reduced carbon emissions, increased biodiversity, and assorted other benefits that result from organic farming.

Traditional organic farming is labor and knowledge-intensive whereas conventional farming is capital-intensive, requiring more energy and manufactured inputs.

Organic farmers in California have cited marketing as their greatest obstacle.

Geographic Producer Distribution

The markets for organic products are strongest in North America and Europe, which as of 2001 are estimated to have $6 and $8 billion respectively of the $20 billion global market. As of 2007 Australasia has 39% of the total organic farmland, including Australia's 1,180,000 hectares (2,900,000 acres) but 97 percent of this land is sprawling rangeland (2007:35). US sales are 20x as much. Europe farms 23 percent of global organic farmland (6,900,000 ha (17,000,000 acres)), followed by Latin America with 19 percent (5.8 million hectares - 14.3 million acres). Asia has 9.5 percent while North America has 7.2 percent. Africa has 3 percent.

Besides Australia, the countries with the most organic farmland are Argentina (3.1 million hectares - 7.7 million acres), China (2.3 million hectares - 5.7 million acres), and the United States (1.6 million hectares - 4 million acres). Much of Argentina's organic farmland is pasture, like that of Australia (2007:42). Spain, Germany, Brazil (the world's largest agricultural exporter), Uruguay, and the UK follow the United States in the amount of organic land (2007:26).

In the European Union (EU25) 3.9% of the total utilized agricultural area was used for organic production in 2005. The countries with the highest proportion of organic land were Austria (11%) and Italy (8.4%), followed by the Czech Republic and Greece (both 7.2%). The lowest figures were shown for Malta (0.1%), Poland (0.6%) and Ireland (0.8%). In 2009, the proportion of organic land in the EU grew to 4.7%. The countries with highest share of agricultural land were Liechtenstein (26.9%), Austria (18.5%) and Sweden (12.6%). 16% of all farmers in Austria produced organically in 2010. By the same year the proportion of organic land increased to 20%.: In 2005 168,000 ha (415,000 ac) of land in Poland was under organic management. In 2012, 288,261 hectares (712,308 acres) were under organic production, and there were about 15,500 organic farmers; retail sales of organic products were EUR 80 million in 2011. As of 2012 organic exports were part of the government's economic development strategy.

After the collapse of the Soviet Union in 1991, agricultural inputs that had previously been purchased from Eastern bloc countries were no longer available in Cuba, and many Cuban farms converted to organic methods out of necessity. Consequently, organic agriculture is a mainstream practice in Cuba, while it remains an alternative practice in most other countries. Cuba's organic strategy includes development of genetically modified crops; specifically corn that is resistant to the palomilla moth

Growth

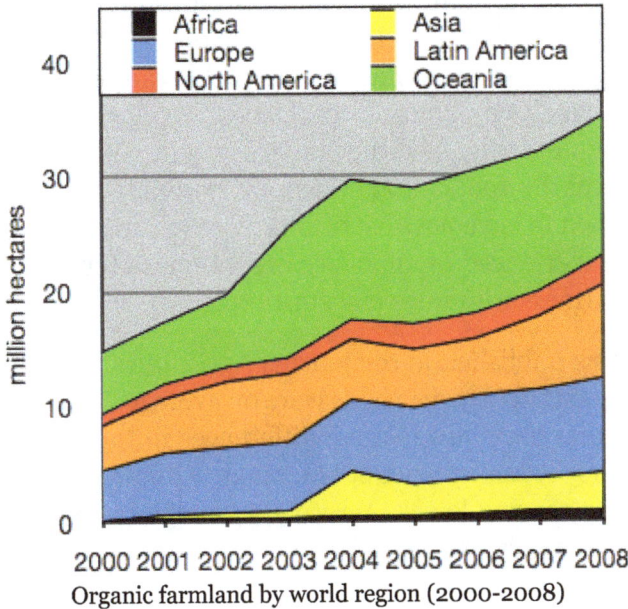

Organic farmland by world region (2000-2008)

In 2001, the global market value of certified organic products was estimated at USD $20 billion. By 2002, this was USD $23 billion and by 2015 more than USD $43 billion. By 2014, retail sales of organic products reached USD $80 billion worldwide. North America and Europe accounted for more than 90% of all organic product sales.

Organic agricultural land increased almost fourfold in 15 years, from 11 million hectares in 1999 to 43.7 million hectares in 2014. Between 2013 and 2014, organic agricultural land grew by 500,000 hectares worldwide, increasing in every region except Latin America. During this time period, Europe's organic farmland increased 260,000 hectares to 11.6 million total (+2.3%), Asia's increased 159,000 hectares to 3.6 million total (+4.7%), Africa's increased 54,000 hectares to 1.3 million total (+4.5%), and North America's increased 35,000 hectares to 3.1 million total (+1.1%). As of 2014, the country with the most organic land was Australia (17.2 million hectares), followed by Argentina (3.1 million hectares), and the United States (2.2 million hectares).

In 2013, the number of organic producers grew by almost 270,000, or more than 13%. By 2014, there were a reported 2.3 million organic producers in the world. Most of the total global increase took place in the Philippines, Peru, China, and Thailand. Overall, the majority of all organic producers are in India (650,000 in 2013), Uganda (190,552 in 2014), Mexico (169,703 in 2013) and the Philippines (165,974 in 2014).

Productivity

Studies comparing yields have had mixed results. These differences among findings can often be attributed to variations between study designs including differences in the crops studied and the methodology by which results were gathered.

A 2012 meta-analysis found that productivity is typically lower for organic farming than conventional farming, but that the size of the difference depends on context and in some cases may be very small. While organic yields can be lower than conventional yields, another meta-analysis published in Sustainable Agriculture Research in 2015, concluded that certain organic on-farm practices could help narrow this gap. Timely weed management and the application of manure in conjunction with legume forages/ cover crops were shown to have positive results in increasing organic corn and soybean productivity. More experienced organic farmers were also found to have higher yields than other organic farmers who were just starting out.

Another meta-analysis published in the journal Agricultural Systems in 2011 analyzed 362 datasets and found that organic yields were on average 80% of conventional yields. The author's found that there are relative differences in this yield gap based on crop type with crops like soybeans and rice scoring higher than the 80% average and crops like wheat and potato scoring lower. Across global regions, Asia and Central Europe were found to have relatively higher yields and Northern Europe relatively lower than the average.

A 2007 study compiling research from 293 different comparisons into a single study to assess the overall efficiency of the two agricultural systems has concluded that "organic methods could produce enough food on a global per capita basis to sustain the current human population, and potentially an even larger population, without increasing the agricultural land base." The researchers also found that while in developed countries, organic systems on average produce 92% of the yield produced by conventional agri-culture, organic systems produce 80% more than conventional farms in developing countries, because the materials needed for organic farming are more accessible than synthetic farming materials to farmers in some poor countries. This study was strongly contested by another study published in 2008, which stated, and was entitled, "Organic agriculture cannot feed the world" and said that the 2007 came up with "a major over-estimation of the productivity of OA" "because data are misinterpreted and calculations accordingly are erroneous." Additional research needs to be conducted in the future to further clarify these claims.

Long Term Studies

A study published in 2005 compared conventional cropping, organic animal-based cropping, and organic legume-based cropping on a test farm at the Rodale Institute over 22 years. The study found that "the crop yields for corn and soybeans were similar in the organic animal, organic legume, and conventional farming systems". It also found that "significantly less fossil energy was expended to produce corn in the Rodale Institute's organic animal and organic legume systems than in the conventional production system. There was little difference in energy input between the different treatments for producing soybeans. In the organic systems, synthetic fertilizers and pesticides were generally not used". As of 2013 the Rodale study was ongoing and a thirty-year anniversary report was published by Rodale in 2012.

A long-term field study comparing organic/conventional agriculture carried out over 21 years in Switzerland concluded that "Crop yields of the organic systems averaged over 21 experimental years at 80% of the conventional ones. The fertilizer input, however, was 34 – 51% lower, indicating an efficient production. The organic farming systems used 20 – 56% less energy to produce a crop unit and per land area this difference was 36 – 53%. In spite of the considerably lower pesticide input the quality of organic products was hardly discernible from conventional analytically and even came off better in food preference trials and picture creating methods"

Profitability

In the United States, organic farming has been shown to be 2.9 to 3.8 times more profitable for the farmer than conventional farming when prevailing price premiums are taken into account. Globally, organic farming is between 22 and 35 percent more profitable for farmers than conventional methods, according to a 2015 meta-analysis of studies conducted across five continents.

The profitability of organic agriculture can be attributed to a number of factors. First, organic farmers do not rely on synthetic fertilizer and pesticide inputs, which can be costly. In addition, organic foods currently enjoy a price premium over conventionally produced foods, meaning that organic farmers can often get more for their yield.

The price premium for organic food is an important factor in the economic viability of organic farming. In 2013 there was a 100% price premium on organic vegetables and a 57% price premium for organic fruits. These percentages are based on wholesale fruit and vegetable prices, available through the United States Department of Agriculture's Economic Research Service. Price premiums exist not only for organic versus nonorganic crops, but may also vary depending on the venue where the product is sold: farmers markets, grocery stores, or wholesale to restaurants. For many producers, direct sales at farmers markets are most profitable because the farmer receives the entire markup, however this is also the most time and labor-intensive approach.

There have been signs of organic price premiums narrowing in recent years, which lowers the economic incentive for farmers to convert to or maintain organic production methods. Data from 22 years of experiments at the Rodale Institute found that, based on the current yields and production costs associated with organic farming in the United States, a price premium of only 10% is required to achieve parity with conventional farming. A separate study found that on a global scale, price premiums of only 5-7% percent were needed to break even with conventional methods. Without the price premium, profitability for farmers is mixed.

For markets and supermarkets organic food is profitable as well, and is generally sold at significantly higher prices than non-organic food.

Energy Efficiency

In the most recent assessments of the energy efficiency of organic versus conventional agriculture, results have been mixed regarding which form is more carbon efficient. Organic farm systems have more often than not been found to be more energy efficient, however, this is not always the case. More than anything, results tend to depend upon crop type and farm size.

A comprehensive comparison of energy efficiency in grain production, produce yield, and animal husbandry concluded that organic farming had a higher yield per unit of energy over the vast majority of the crops and livestock systems. For example, two studies - both comparing organically- versus conventionally-farmed apples - declare contradicting results, one saying organic farming is more energy efficient, the other saying conventionally is more efficient.

It has generally been found that the labor input per unit of yield was higher for organic systems compared with conventional production.

Sales and Marketing

Most sales are concentrated in developed nations. In 2008, 69% of Americans claimed to occasionally buy organic products, down from 73% in 2005. One theory for this change was that consumers were substituting "local" produce for "organic" produce.

Distributors

The USDA requires that distributors, manufacturers, and processors of organic products be certified by an accredited state or private agency. In 2007, there were 3,225 certified organic handlers, up from 2,790 in 2004.

Organic handlers are often small firms; 48% reported sales below $1 million annually, and 22% between $1 and $5 million per year. Smaller handlers are more likely to sell to independent natural grocery stores and natural product chains whereas large distributors more often market to natural product chains and conventional supermarkets, with a small group marketing to independent natural product stores. Some handlers work with conventional farmers to convert their land to organic with the knowledge that the farmer will have a secure sales outlet. This lowers the risk for the handler as well as the farmer. In 2004, 31% of handlers provided technical support on organic standards or production to their suppliers and 34% encouraged their suppliers to transition to organic. Smaller farms often join together in cooperatives to market their goods more effectively.

93% of organic sales are through conventional and natural food supermarkets and chains, while the remaining 7% of U.S. organic food sales occur through farmers' markets, foodservices, and other marketing channels.

Direct-to-consumer Sales

In the 2012 Census, direct-to-consumer sales equaled $1.3 billion, up from $812 million in 2002, an increase of 60 percent. The number of farms that utilize direct-to-consumer sales was 144,530 in 2012 in comparison to 116,733 in 2002. Direct-to-consumer sales include farmers markets, community supported agriculture (CSA), on-farm stores, and roadside farm stands. Some organic farms also sell products direct to retailer, direct to restaurant and direct to institution. According to the 2008 Organic Production Survey, approximately 7% of organic farm sales went direct-to-consumers, 10% went direct to retailers, and approximately 83% went into wholesale markets. In comparison, only 0.4% of the value of convention agricultural commodities went direct-to-consumers.

While not all products sold at farmer's markets are certified organic, this direct-to-consumer avenue has become increasingly popular in local food distribution and has grown substantially since 1994. In 2014, there were 8,284 farmer's markets in comparison to 3,706 in 2004 and 1,755 in 1994, most of which are found in populated areas such as the Northeast, Midwest, and West Coast.

Labor and Employment

Organic production is more labor-intensive than conventional production. On the one hand, this increased labor cost is one factor that makes organic food more expensive. On the other hand, the increased need for labor may be seen as an "employment dividend" of organic farming, providing more jobs per unit area than conventional systems. The 2011 UNEP Green Economy Report suggests that "[a]n increase in investment in green agriculture is projected to lead to growth in employment of about 60 per cent compared with current levels" and that "green agriculture investments could create 47 million additional jobs compared with BAU2 over the next 40 years." The UNEP also argues that "[b]y greening agriculture and food distribution, more calories per person per day, more jobs and business opportunities especially in rural areas, and market-access opportunities, especially for developing countries, will be available."

World's Food Security

In 2007 the United Nations Food and Agriculture Organization (FAO) said that organic agriculture often leads to higher prices and hence a better income for farmers, so it should be promoted. However, FAO stressed that by organic farming one could not feed the current mankind, even less the bigger future population. Both data and models showed then that organic farming was far from sufficient. Therefore, chemical fertilizers were needed to avoid hunger. Other analysis by many agribusiness executives, agricultural and ecological scientists, and international agriculture experts revealed the opinion that organic farming would not only increase the world's food supply, but might be the only way to eradicate hunger.

FAO stressed that fertilizers and other chemical inputs can much increase the production, particularly in Africa where fertilizers are currently used 90% less than in Asia. For example, in Malawi the yield has been boosted using seeds and fertilizers. FAO also calls for using biotechnology, as it can help smallholder farmers to improve their income and food security.

Also NEPAD, development organization of African governments, announced that feeding Africans and preventing malnutrition requires fertilizers and enhanced seeds.

According to a more recent study in ScienceDigest, organic best management practices shows an average yield only 13% less than conventional. In the world's poorer nations where most of the world's hungry live, and where conventional agriculture's expensive inputs are not affordable by the majority of farmers, adopting organic management actually increases yields 93% on average, and could be an important part of increased food security.

Capacity Building in Developing Countries

Organic agriculture can contribute to ecologically sustainable, socio-economic development, especially in poorer countries. The application of organic principles enables employment of local resources (e.g., local seed varieties, manure, etc.) and therefore cost-effectiveness. Local and international markets for organic products show tremendous growth prospects and offer creative producers and exporters excellent opportunities to improve their income and living conditions.

Organic agriculture is knowledge intensive. Globally, capacity building efforts are underway, including localized training material, to limited effect. As of 2007, the International Federation of Organic Agriculture Movements hosted more than 170 free manuals and 75 training opportunities online.

In 2008 the United Nations Environmental Programme (UNEP) and the United Nations Conference on Trade and Development (UNCTAD) stated that "organic agriculture can be more conducive to food security in Africa than most conventional production systems, and that it is more likely to be sustainable in the long-term" and that "yields had more than doubled where organic, or near-organic practices had been used" and that soil fertility and drought resistance improved.

Millennium Development Goals

The value of organic agriculture (OA) in the achievement of the Millennium Development Goals (MDG), particularly in poverty reduction efforts in the face of climate change, is shown by its contribution to both income and non-income aspects of the MDGs. These benefits are expected to continue in the post-MDG era. A series of case studies conducted in selected areas in Asian countries by the Asian Development Bank Institute (ADBI) and published as a book compilation by ADB in Manila document these contributions to both income and non-income aspects of the MDGs. These include poverty alleviation by way of higher incomes, improved farmers' health owing to less chemical exposure, integration of sustainable principles into rural development policies, improvement of access to safe water and sanitation, and expansion of global partnership for development as small farmers are integrated in value chains.

A related ADBI study also sheds on the costs of OA programs and set them in the context of the costs of attaining the MDGs. The results show considerable variation across the case studies, suggesting that there is no clear structure to the costs of adopting OA. Costs depend on the efficiency of the OA adoption programs. The lowest cost programs were more than ten times less expensive than the highest cost ones. However, further analysis of the gains resulting from OA adoption reveals that the costs per person taken out of poverty was much lower than the estimates of the World Bank, based on income growth in general or based on the detailed costs of meeting some of the more quantifiable MDGs (e.g., education, health, and environment).

Externalities

Agriculture imposes negative externalities (uncompensated costs) upon society through public land and other public resource use, biodiversity loss, erosion, pesticides, nutrient runoff, subsidized water usage, subsidy payments and assorted other problems. Positive externalities include self-reliance, entrepreneurship, respect for nature, and air quality. Organic methods reduce some of these costs. In 2000 uncompensated costs for 1996 reached 2,343 million British pounds or £208 per ha (£84.20/ac). A study of practices in the USA published in 2005 concluded that cropland costs the economy approximately 5 to 16 billion dollars ($30–96/ha - $12–39/ac), while livestock production costs 714 million dollars. Both studies recommended reducing externalities. The

2000 review included reported pesticide poisonings but did not include speculative chronic health effects of pesticides, and the 2004 review relied on a 1992 estimate of the total impact of pesticides.

It has been proposed that organic agriculture can reduce the level of some negative externalities from (conventional) agriculture. Whether the benefits are private or public depends upon the division of property rights.

Several surveys and studies have attempted to examine and compare conventional and organic systems of farming and have found that organic techniques, while not without harm, are less damaging than conventional ones because they reduce levels of biodiversity less than conventional systems do and use less energy and produce less waste when calculated per unit area.

A 2003 to 2005 investigation by the Cranfield University for the Department for Environment Food and Rural Affairs in the UK found that it is difficult to compare the Global Warming Potential (GWP), acidification and eutrophication emissions but "Organic production often results in increased burdens, from factors such as N leaching and N2O emissions", even though primary energy use was less for most organic products. N_2O is always the largest GWP contributor except in tomatoes. However, "organic tomatoes always incur more burdens (except pesticide use)". Some emissions were lower "per area", but organic farming always required 65 to 200% more field area than non-organic farming. The numbers were highest for bread wheat (200+ % more) and potatoes (160% more).

The situation was shown dramatically in a comparison of a modern dairy farm in Wisconsin with one in New Zealand in which the animals grazed extensively. Using total farm emissions per kg milk produced as a parameter, the researchers showed that production of methane from belching was higher in the New Zealand farm, while carbon dioxide production was higher in the Wisconsin farm. Output of nitrous oxide, a gas with an estimated global warming potential 310 times that of carbon dioxide was also higher in the New Zealand farm. Methane from manure handling was similar in the two types of farm. The explanation for the finding relates to the different diets used on these farms, being based more completely on forage (and hence more fibrous) in New Zealand and containing less concentrate than in Wisconsin. Fibrous diets promote a higher proportion of acetate in the gut of ruminant animals, resulting in a higher production of methane that must be released by belching. When cattle are given a diet containing some concentrates (such as corn and soybean meal) in addition to grass and silage, the pattern of ruminal fermentation alters from acetate to mainly propionate. As a result, methane production is reduced. Capper et al. compared the environmental impact of US dairy production in 1944 and 2007. They calculated that the carbon "footprint" per billion kg (2.2 billion lb) of milk produced in 2007 was 37 percent that of equivalent milk production in 1944.

Environmental Impact and Emissions

Researchers at Oxford university analyzed 71 peer-reviewed studies and observed that organic products are sometimes worse for the environment. Organic milk, cereals, and pork generated higher greenhouse gas emissions per product than conventional ones but organic beef and olives had lower emissions in most studies. Usually organic products required less energy, but more land. Per unit of product, organic produce generates higher nitrogen leaching, nitrous oxide emissions, ammonia emissions, eutrophication and acidification potential than when conventionally grown. Other differences were not significant. The researchers concluded, as there is not singular way of doing conventional or organic farming, that the debate should go beyond the conventional vs organic debate, and more about finding specific solutions to specific circumstances.

Proponents of organic farming have claimed that organic agriculture emphasizes closed nutrient cycles, biodiversity, and effective soil management providing the capacity to mitigate and even reverse the effects of climate change and that organic agriculture can decrease fossil fuel emissions. "The carbon sequestration efficiency of organic systems in temperate climates is almost double (575-700 kg carbon per ha per year - 510-625 lb/ac/an) that of conventional treatment of soils, mainly owing to the use of grass clovers for feed and of cover crops in organic rotations."

Critics of organic farming methods believe that the increased land needed to farm organic food could potentially destroy the rainforests and wipe out many ecosystems.

Nutrient Leaching

According to the meta-analysis of 71 studies, nitrogen leaching, nitrous oxide emissions, ammonia emissions, eutrophication potential and acidification potential were higher for organic products, although in one study "nitrate leaching was 4.4-5.6 times higher in conventional plots than organic plots".

Excess nutrients in lakes, rivers, and groundwater can cause algal blooms, eutrophication, and subsequent dead zones. In addition, nitrates are harmful to aquatic organisms by themselves.

Land Use

The Oxford meta-analysis of 71 studies proved that organic farming requires 84% more land, mainly due to lack of nutrients but sometimes due to weeds, diseases or pests, lower yielding animals and land required for fertility building crops. While organic farming does not necessarily save land for wildlife habitats and forestry in all cases, the most modern breakthroughs in organic are addressing these issues with success.

Professor Wolfgang Branscheid says that organic animal production is not good for the environment, because organic chicken requires doubly as much land as "conventional" chicken and organic pork a quarter more. According to a calculation by Hudson Institute, organic beef requires triply as much land. On the other hand, certain organic methods of animal husbandry have been shown to restore desertified, marginal, and/or otherwise unavailable land to agricultural productivity and wildlife. Or by getting both forage and cash crop production from the same fields simultaneously, reduce net land use.

In England organic farming yields 55% of normal yields. While in other regions of the world, organic methods have started producing record yields.

Pesticides

A sign outside of an organic apple orchard in Pateros, Washington reminding orchardists not to spray pesticides on these trees.

In organic farming synthetic pesticides are generally prohibited. A chemical is said to be synthetic if it does not already exist in the natural world. But the organic label goes further and usually prohibit compounds that exist in nature if they are produced by Chemical synthesis. So the prohibition is also about the method of production and not only the nature of the compound.

An non exhaustive list of organic approved pesticides with theirs Median lethal dose

- Copper(II) sulfate is used as a fungicide and is also used in conventional agriculture (LD_{50} 300 mg/kg). Conventional agriculture has the option to use the less toxic Mancozeb (LD_{50} 4,500 to 11,200 mg/kg)

- Boric acid is used as stomach poison that target insects (LD_{50}: 2660 mg/kg).

- Pyrethrin comes from chemicals extracted from flowers of the genus Pyrethrum (LD_{50} of 370 mg/kg). Its potent toxicity is used to control insects.

- Lime sulphur (aka calcium polysulfide) and sulfur are considered to be allowed, synthetic materials (LD_{50}: 820 mg/kg)

- Rotenone is a powerful insecticide that was used to control insects (LD_{50}: 132 mg/kg). Despite the high toxicity of Rotenone to aquatic life and some links to Parkinson disease the compound is still allowed in organic farming as it is a naturally occurring compound.

- Bromomethane is a gas that is still used in the nurseries of Strawberry organic farming

- Azadirachtin is a wide spectrum very potent insecticide. Almost non toxic to mammals (LD_{50} in rats is > 3,540 mg/kg) but affects beneficial insects.

Food Quality and safety

While there may be some differences in the amounts of nutrients and anti-nutrients when organically produced food and conventionally produced food are compared, the variable nature of food production and handling makes it difficult to generalize results, and there is insufficient evidence to make claims that organic food is safer or healthier than conventional food. Claims that organic food tastes better are not supported by evidence.

Soil Conservation

Supporters claim that organically managed soil has a higher quality and higher water retention. This may help increase yields for organic farms in drought years. Organic farming can build up soil organic matter better than conventional no-till farming, which suggests long-term yield benefits from organic farming. An 18-year study of organic methods on nutrient-depleted soil concluded that conventional methods were superior for soil fertility and yield for nutrient-depleted soils in cold-temperate climates, arguing that much of the benefit from organic farming derives from imported materials that could not be regarded as self-sustaining.

In *Dirt: The Erosion of Civilizations*, geomorphologist David Montgomery outlines a coming crisis from soil erosion. Agriculture relies on roughly one meter of topsoil, and that is being depleted ten times faster than it is being replaced. No-till farming, which some claim depends upon pesticides, is one way to minimize erosion. However, a 2007 study by the USDA's Agricultural Research Service has found that manure applications in tilled organic farming are better at building up the soil than no-till.

Biodiversity

The conservation of natural resources and biodiversity is a core principle of organic production. Three broad management practices (prohibition/reduced use of chemical pesticides and inorganic fertilizers; sympathetic management of non-cropped habitats; and preservation of mixed farming) that are largely intrinsic (but not exclusive) to organic farming are particularly beneficial for farmland wildlife. Using practices that attract or introduce beneficial insects, provide habitat for birds and mammals, and provide conditions that increase soil biotic diversity serve to supply vital ecological services to organic production systems. Advantages to certified organic operations that implement these types of production practices include: 1) decreased dependence on outside fertility inputs; 2) reduced pest management costs; 3) more reliable sources of clean water; and 4) better pollination.

Nearly all non-crop, naturally occurring species observed in comparative farm land practice studies show a preference for organic farming both by abundance and diversity. An average of 30% more species inhabit organic farms. Birds, butterflies, soil microbes, beetles, earthworms, spiders, vegetation, and mammals are particularly affected. Lack of herbicides and pesticides improve biodiversity fitness and population density. Many weed species attract beneficial insects that improve soil qualities and forage on weed pests. Soil-bound organisms often benefit because of increased bacteria populations due to natural fertilizer such as manure, while experiencing reduced intake of herbicides and pesticides. Increased biodiversity, especially from beneficial soil microbes and mycorrhizae have been proposed as an explanation for the high yields experienced by some organic plots, especially in light of the differences seen in a 21-year comparison of organic and control fields.

Biodiversity from organic farming provides capital to humans. Species found in organic farms enhance sustainability by reducing human input (e.g., fertilizers, pesticides).

The USDA's Agricultural Marketing Service (AMS) published a *Federal Register* notice on 15 January 2016, announcing the National Organic Program (NOP) final guidance on Natural Resources and Biodiversity Conservation for Certified Organic Operations. Given the broad scope of natural resources which includes soil, water, wetland, woodland and wildlife, the guidance provides examples of practices that support the underlying conservation principles and demonstrate compliance with USDA organic regulations § 205.200. The final guidance provides organic certifiers and farms with examples of production practices that support conservation principles and comply with the USDA organic regulations, which require operations to maintain or improve natural resources. The final guidance also clarifies the role of certified operations (to submit an OSP to a certifier), certifiers (ensure that the OSP describes or lists practices that explain the operator's monitoring plan and practices to support natural resources and biodiversity conservation), and inspectors (onsite inspection) in the implementation and verification of these production practices.

A wide range of organisms benefit from organic farming, but it is unclear whether organic methods confer greater benefits than conventional integrated agri-environmental programs. Organic farming is often presented as a more biodiversity-friendly practice, but the generality of the beneficial effects of organic farming is debated as the effects appear often species- and context-dependent, and current research has highlighted the need to quantify the relative effects of local- and landscape-scale management on farmland biodiversity. There are four key issues when comparing the impacts on biodiversity of organic and conventional farming: (1) It remains unclear whether a holistic whole-farm approach (i.e. organic) provides greater benefits to biodiversity than carefully targeted prescriptions applied to relatively small areas of cropped and/or non-cropped habitats within conventional agriculture (i.e. agri-environment schemes); (2) Many comparative studies encounter methodological problems, limiting their ability to draw quantitative conclusions; (3) Our knowledge of the impacts of organic farming in pastoral and upland agriculture is limited; (4) There remains a pressing need for longitudinal, system-level studies in order to address these issues and to fill in the gaps in our knowledge of the impacts of organic farming, before a full appraisal of its potential role in biodiversity conservation in agroecosystems can be made.

Regional Support for Organic Farming

India

In India, states such as Sikkim and Kerala have planned to shift to fully organic cultivation by 2015 and 2016 respectively.

Organic Food

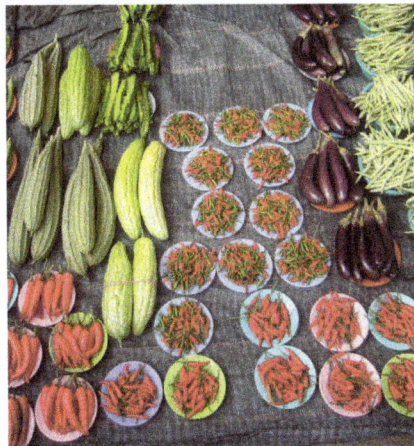

Organic vegetables at a farmers' market in Argentina

Organic foods are foods produced by methods that comply with the standards of organic farming. Standards vary worldwide; however, organic farming in general, features practices that strive to foster cycling of resources, promote ecological balance, and conserve biodiversity. Organizations regulating organic products may choose to restrict the use of certain pesticides and fertilizers in farming. In general, organic foods are also usually not processed using irradiation, industrial solvents or synthetic food additives.

Currently, the European Union, the United States, Canada, Mexico, Japan, and many other countries require producers to obtain special certification in order to market food as organic, within their borders. In the context of these regulations, organic food is food produced in a way that complies with organic standards set by national governments and international organizations. Although the produce of kitchen gardens may be organic, selling food with the organic label is regulated by governmental food safety authorities, such as the US Department of Agriculture (USDA) or European Commission.

There is no sufficient evidence in medical literature to support claims that organic food is safer or healthier than conventionally grown food. While there may be some differences in the nutrient and anti-nutrient contents of organically and conventionally produced food, the variable nature of food production and handling makes it difficult to generalize results. Claims that organic food tastes better are generally not supported by evidence.

Meaning and Origin of the Term

For the vast majority of its history, agriculture can be described as having been organic; only during the 20th century was a large supply of new products, generally deemed not organic, introduced into food production. The organic farming movement arose in the 1940s in response to the industrialization of agriculture.

In 1939, Lord Northbourne coined the term *organic farming* in his book *Look to the Land* (1940), out of his conception of "the farm as organism," to describe a holistic, ecologically balanced approach to farming—in contrast to what he called *chemical farming*, which relied on "imported fertility" and "cannot be self-sufficient nor an organic whole." Early soil scientists also described the differences in soil composition when animal manures were used as "organic", because they contain carbon compounds where superphosphates and haber process nitrogen do not. Their respective use affects humus content of soil. This is different from the scientific use of the term "organic" in chemistry, which refers to a class of molecules that contain carbon, especially those involved in the chemistry of life. This class of molecules includes everything likely to be considered edible, and include most pesticides and toxins too, therefore the term "organic" and, especially, the term "inorganic" (sometimes wrongly used as a contrast by the popular press) as they apply to organic chemistry is an equivocation fallacy when applied to farming, the production of food, and to foodstuffs themselves. Properly used in this agricultural science context, "organic" refers to the methods grown and processed, not necessarily the chemical composition of the food.

Ideas that organic food could be healthier and better for the environment originated in the early days of the organic movement as a result of publications like the 1943 book The Living Soil and *Farming and Gardening for Health or Disease* (1945).

Early consumers interested in organic food would look for non-chemically treated, non-use of unapproved pesticides, fresh or minimally processed food. They mostly had to buy directly from growers. Later, "Know your farmer, know your food" became the motto of a new initiative instituted by the USDA in September 2009. Personal definitions of what constituted "organic" were developed through firsthand experience: by talking to farmers, seeing farm conditions, and farming activities. Small farms grew vegetables (and raised livestock) using organic farming practices, with or without certification, and the individual consumer monitored. Small specialty health food stores and co-operatives were instrumental to bringing organic food to a wider audience. As demand for organic foods continued to increase, high volume sales through mass outlets such as supermarkets rapidly replaced the direct farmer connection. Today, many large corporate farms have an organic division. However, for supermarket consumers, food production is not easily observable, and product labeling, like "certified organic", is relied upon. Government regulations and third-party inspectors are looked to for assurance.

In the 1970s, interest in organic food grew with the publication of Silent Spring and the rise of the environmental movement, and was also spurred by food-related health scares like the concerns about Alar that arose in the mid-1980s.

Legal Definition

The National Organic Program (run by the USDA) is in charge of the legal definition of *organic* in the United States and does organic certification.

Organic food production is a self-regulated industry with government oversight in some countries, distinct from private gardening. Currently, the European Union, the United States, Canada, Japan, and many other countries require producers to obtain special certification based on government-defined standards in order to market food as organic within their borders. In the context of these regulations, foods marketed as organic are produced in a way that complies with organic standards set by national governments and international organic industry trade organizations.

In the United States, organic production is managed in accordance with the Organic Foods Production Act of 1990 (OFPA) and regulations in Title 7, Part 205 of the Code of Federal Regulations to respond to site-specific conditions by integrating cultural, biological, and mechanical practices that foster cycling of resources, promote ecological balance, and conserve biodiversity. If livestock are involved, the livestock must be reared with regular access to pasture and without the routine use of antibiotics or growth hormones.

Processed organic food usually contains only organic ingredients. If non-organic ingredients are present, at least a certain percentage of the food's total plant and animal ingredients must be organic (95% in the United States, Canada, and Australia). Foods claiming to be organic must be free of artificial food additives, and are often processed with fewer artificial methods, materials and conditions, such as chemical ripening, food irradiation, and genetically modified ingredients. Pesticides are allowed as long as they are not synthetic. However, under US federal organic standards, if pests and weeds are not controllable through management practices, nor via organic pesticides and herbicides, "a substance included on the National List of synthetic substances allowed for use in organic crop production may be applied to prevent, suppress, or control pests, weeds, or diseases." Several groups have called for organic standards to prohibit nanotechnology on the basis of the precautionary principle in light of unknown risks of nanotechnology. The use of nanotechnology-based products in the production of organic food is prohibited in some jurisdictions (Canada, the UK, and Australia) and is unregulated in others.

To be certified organic, products must be grown and manufactured in a manner that adheres to standards set by the country they are sold in:

- Australia: NASAA Organic Standard

- Canada:

- European Union: EU-Eco-regulation

    o Sweden: KRAV

    o United Kingdom: DEFRA

    o Poland: Association of Polish Ecology

    o Norway: Debio Organic certification

- India: NPOP, (National Program for Organic Production)

- Indonesia: BIOCert, run by Agricultural Ministry of Indonesia.

- Japan: JAS Standards

- United States: National Organic Program (NOP) Standards

In the United States, there are four different levels or categories for organic labeling. 1)'100%' Organic: This means that all ingredients are produced organically. It also may have the USDA seal. 2)'Organic': At least 95% or more of the ingredients are organic. 3)'Made With Organic Ingredients': Contains at least 70% organic ingredients. 4)'Less Than 70% Organic Ingredients': Three of the organic ingredients must be listed under the ingredient section of the label. In the U.S., the food label "natural" or "all natural" does not mean that the food was produced and processed organically.

Public Perception

There is widespread public belief that organic food is safer, more nutritious, and better tasting than conventional food. Consumers purchase organic foods for different reasons, including concerns about the effects of conventional farming practices on the environment, human health, and animal welfare.

The most important reason for purchasing organic foods seems to be beliefs about the products' health-giving properties and higher nutritional value. These beliefs are promoted by the organic food industry, and have fueled increased demand for organic food despite higher prices and difficulty in confirming these claimed benefits scientifically. Organic labels also stimulate the consumer to view the product as having more positive nutritional value.

Psychological effects such as the "halo" effect, which are related to the choice and consumption of organic food, are also important motivating factors in the purchase of organic food. The perception that organic food is low-calorie food or health food appears to be common.

In China the increasing demand for organic products of all kinds, and in particular milk, baby food and infant formula, has been "spurred by a series of food scares, the worst being the death of six children who had consumed baby formula laced with melamine" in 2009 and the 2008 Chinese milk scandal, making the Chinese market for organic milk the largest in the world as of 2014. A Pew Research Centre survey in 2012 indicated that 41% of Chinese consumers thought of food safety as a very big problem, up by three times from 12% in 2008.

Taste

There is no good evidence that organic food tastes better than its non-organic counterparts. There is evidence that some organic fruit is drier than conventionally grown fruit; a slightly drier fruit may also have a more intense flavor due to the higher concentration of flavoring substances.

Some foods, such as bananas, are picked when unripe, are cooled to prevent ripening while they are shipped to market, and then are induced to ripen quickly by exposing them to propylene or ethylene, chemicals produced by plants to induce their own ripening; as flavor and texture changes during ripening, this process may affect those qualities of the treated fruit. The issue of ethylene use to ripen fruit in organic food production is contentious because ripeness when picked often does affect taste; opponents claim that its use benefits only large companies and that it opens the door to weaker organic standards.

Chemical Composition

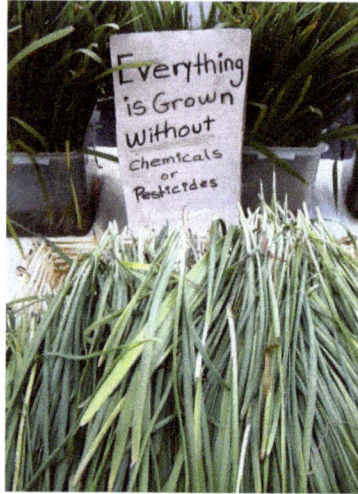

Organic vegetables at a farmers' market

With respect to chemical differences in the composition of organically grown food compared with conventionally grown food, studies have examined differences in nutrients, antinutrients, and pesticide residues. These studies generally suffer from confounding variables, and are difficult to generalize due to differences in the tests that were done, the methods of testing, and because the vagaries of agriculture affect the chemical composition of food; these variables include variations in weather (season to season as well as place to place); crop treatments (fertilizer, pesticide, etc.); soil composition; the cultivar used, and in the case of meat and dairy products, the parallel variables in animal production. Treatment of the foodstuffs after initial gathering (whether milk is pasteurized or raw), the length of time between harvest and analysis, as well as conditions of transport and storage, also affect the chemical composition of a given item of food. Additionally, there is evidence that organic produce is drier than conventionally grown produce; a higher content in any chemical category may be explained by higher concentration rather than in absolute amounts.

Nutrients

Many people believe that organic foods have higher content of nutrients and thus are healthier than conventionally produced foods. However, scientists have not been equally convinced that this is the case as the research conducted in the field has not shown consistent results.

A 2009 systematic review in the American Journal of Clinical Nutrition found that organically produced foodstuffs are not richer in vitamins and minerals than conventionally produced foodstuffs. The results of the systematic review only showed a lower nitrogen and higher phosphorus content in organic produced compared to conventionally grown foodstuffs. Content of vitamin C, calcium, potassium, total soluble solids, copper, iron, nitrates, manganese, and sodium did not differ between the two categories.

A 2014 meta-analysis of 343 studies found that organically grown crops had 17% higher concentrations of polyphenols than conventionally grown crops. Concentrations of phenolic acids, flavanones, stilbenes, flavones, flavonols, and anthocyanins were elevated, with flavanones being 69% higher.

A 2012 survey of the scientific literature did not find significant differences in the vitamin content of organic and conventional plant or animal products, and found that results varied from study to study. Produce studies reported on ascorbic acid (Vitamin C) (31 studies), beta-carotene (a precursor for Vitamin A) (12 studies), and alpha-tocopherol (a form of Vitamin E) (5 studies) content; milk studies reported on beta-carotene (4 studies) and alpha-tocopherol levels (4 studies). Few studies examined vitamin content in meats, but these found no difference in beta-carotene in beef, alpha-tocopherol in pork or beef, or vitamin A (retinol) in beef. The authors analyzed 11 other nutrients reported in studies of produce. Only two nutrients were significantly higher in organic than conventional produce: phosphorus and total polyphenols). A 2011 literature review found that organic foods had a higher micronutrient content overall than conventionally produced foods.

Similarly, organic chicken contained higher levels of omega-3 fatty acids than conventional chicken. The authors found no difference in the protein or fat content of organic and conventional raw milk.

A 2016 systematic review and meta-analysis found that organic meat had comparable or slightly lower levels of saturated fat and monounsaturated fat as conventional meat, but higher levels of both overall and n-3 polyunsaturated fatty acids. Another meta-analysis published the same year found no significant differences in levels of saturated and monounsaturated fat between organic and conventional milk, but significantly higher levels of overall and n-3 polyunsaturated fatty acids in organic milk than in conventional milk.

Anti-nutrients

The amount of nitrogen content in certain vegetables, especially green leafy vegetables and tubers, has been found to be lower when grown organically as compared to conventionally. When evaluating environmental toxins such as heavy metals, the USDA has noted that organically raised chicken may have lower arsenic levels. Early literature reviews found no significant evidence that levels of arsenic, cadmium or other heavy metals differed significantly between organic and conventional food products. However, a 2014 review found lower concentrations of cadmium, particularly in organically grown grains.

Pesticide Residues

The amount of pesticides that remain in or on food is called pesticides residue. In the United States, before a pesticide can be used on a food crop, the U.S. Environmental Protection Agency must determine whether that pesticide can be used without posing a risk to human health.

A 2012 meta-analysis determined that detectable pesticide residues were found in 7% of organic produce samples and 38% of conventional produce samples. This result was statistically heterogeneous, potentially because of the variable level of detection used among these studies. Only three studies reported the prevalence of contamination exceeding maximum allowed limits; all were from the European Union. A 2014 meta-analysis found that conventionally grown produce was four times more likely to have pesticide residue than organically grown crops.

The American Cancer Society has stated that no evidence exists that the small amount of pesticide residue found on conventional foods will increase the risk of cancer, though it recommends thoroughly washing fruits and vegetables. They have also stated that there is no research to show that organic food reduces cancer risk compared to foods grown with conventional farming methods.

The Environmental Protection Agency maintains strict guidelines on the regulation of pesticides by setting a tolerance on the amount of pesticide residue allowed to be in or on any particular food. Although some residue may remain at the time of harvest, residue tend to decline as the pesticide breaks down over time. In addition, as the commodities are washed and processed prior to sale, the residues often diminish further.

Bacterial Contamination

A 2012 meta-analysis determined that prevalence of *E. coli* contamination was not statistically significant (7% in organic produce and 6% in conventional produce). While bacterial contamination is common among both organic and conventional animal products, differences in the prevalence of bacterial contamination between organic and conventional animal products were also statistically insignificant.

Organic Meat Production Requirements

United States

Organic meat certification in the United States requires farm animals to be raised according to USDA organic regulations throughout their lives. These regulations require that livestock are fed certified organic food that contains no animal byproducts. Further, organic farm animals can receive no growth hormones or antibiotics, and they must be raised using techniques that protect native species and other natural resources. Irradiation and genetic engineering are not allowed with organic animal production. One of the major differences in organic animal husbandry protocol is the "pasture rule": minimum requirements for time on pasture do vary somewhat by species and between the certifying agencies, but the common theme is to require as much time on pasture as possible and reasonable.

Health and Safety

There is little scientific evidence of benefit or harm to human health from a diet high in organic food, and conducting any sort of rigorous experiment on the subject is very difficult. A 2012 meta-analysis noted that "there have been no long-term studies of health outcomes of populations consuming predominantly organic versus conventionally produced food controlling for socioeconomic factors; such studies would be expensive to conduct." A 2009 meta-analysis noted that "most of the included articles did not study direct human health outcomes. In ten of the included studies (83%), a primary outcome was the change in antioxidant activity. Antioxidant status and activity are useful biomarkers but do not directly equate to a health outcome. Of the remaining two articles, one recorded proxy-reported measures of atopic manifestations as its primary health outcome, whereas the other article examined the fatty acid composition of breast milk and implied possible health benefits for infants from the consumption of different amounts of conjugated linoleic acids from breast milk." In addition, as discussed above, difficulties in accurately and meaningfully measuring chemical differences between organic and conventional food make it difficult to extrapolate health recommendations based solely on chemical analysis.

With regard to the possibility that some organic food may have higher levels of certain anti-oxidants, evidence regarding whether increased anti-oxidant consumption improves health is conflicting.

As of 2012, the scientific consensus is that while "consumers may choose to buy organic fruit, vegetables and meat because they believe them to be more nutritious than other food…. the balance of current scientific evidence does not support this view." A 12-month systematic review commissioned by the FSA in 2009 and conducted at the London School of Hygiene & Tropical Medicine based on 50 years' worth of collected evidence concluded that "there is no good evidence that consumption of organic food is beneficial to health in relation to nutrient content." There is no support in the scientific literature that the lower levels of nitrogen in certain organic vegetables translates to improved health risk.

Consumer Safety

Pesticide Exposure

The main difference between organic and conventional food products are the chemicals involved during production and processing. The residues of those chemicals in food products have dubious effects on the human health. All food products on the market including those that contain residues of pesticides, antibiotics, growth hormones and other types of chemicals that are used during production and processing are said to be safe.

Claims of improved safety of organic food has largely focused on pesticide residues. These concerns are driven by the facts that "(1) acute, massive exposure to pesticides can cause significant adverse health effects; (2) food products have occasionally been contaminated with pesticides, which can result in acute toxicity; and (3) most, if not all, commercially purchased food contains trace amounts of agricultural pesticides." However, as is frequently noted in the scientific literature: "What does not follow from this, however, is that chronic exposure to the trace amounts of pesticides found in food results in demonstrable toxicity. This possibility is practically impossible to study and quantify;" therefore firm conclusions about the relative safety of organic foods have been hampered by the difficulty in proper study design and relatively small number of studies directly comparing organic food to conventional food.

Additionally, the Carcinogenic Potency Project, which is a part of the US EPA's Distributed Structure-Searchable Toxicity (DSSTox) Database Network, has been systemically testing the carcinogenicity of chemicals, both natural and synthetic, and building a publicly available database of the results for the past ~30 years. Their work attempts to fill in the gaps in our scientific knowledge of the carcinogenicity of all chemicals, both natural and synthetic, as the scientists conducting the Project described in the journal, *Science*, in 1992:

Toxicological examination of synthetic chemicals, without similar examination of chemicals that occur naturally, has resulted in an imbalance in both the data on and the perception of chemical carcinogens. Three points that we have discussed indicate that comparisons should be made with natural as well as synthetic chemicals.

1) The vast proportion of chemicals that humans are exposed to occur naturally. Nevertheless, the public tends to view chemicals as only synthetic and to think of synthetic chemicals as toxic despite the fact that every natural chemical is also toxic at some dose. The daily average exposure of Americans to burnt material in the diet is ~2000 mg, and exposure to natural pesticides (the chemicals that plants produce to defend themselves) is ~1500 mg. In comparison, the total daily exposure to all synthetic pesticide residues combined is ~0.09 mg. Thus, we estimate that 99.99% of the pesticides humans ingest are natural. Despite this enormously greater exposure to natural chemicals, 79% (378 out of 479) of the chemicals tested for carcinogenicity in both rats and mice are synthetic (that is, do not occur naturally). 2) It has often been wrongly assumed that humans have evolved defenses against the natural chemicals in our diet but not against the synthetic chemicals. However, defenses that animals have evolved are mostly general rather than specific for particular chemicals; moreover, defenses are generally inducible and therefore protect well from low doses of both synthetic and natural chemicals.

3) Because the toxicology of natural and synthetic chemicals is similar, one expects (and finds) a similar positivity rate for carcinogenicity among synthetic and natural chemicals. The positivity rate among chemicals tested in rats and mice is ~50%. Therefore, because humans are exposed to so many more natural than synthetic chemicals (by weight and by number), humans are exposed to an enormous background of rodent carcinogens, as defined by high-dose tests on rodents. We have shown that even though only a tiny proportion of natural pesticides in plant foods have been tested, the 29 that are rodent carcinogens among the 57 tested, occur in more than 50 common plant foods. It is probable that almost every fruit and vegetable in the supermarket contains natural pesticides that are rodent carcinogens.

While studies have shown via chemical analysis, as discussed above, that organically grown fruits and vegetables have significantly lower pesticide residue levels, the significance of this finding on actual health risk reduction is debatable as both conventional foods and organic foods generally have pesticide levels well below government established guidelines for what is considered safe. This view has been echoed by the U.S. Department of Agriculture and the UK Food Standards Agency.

A study published by the National Research Council in 1993 determined that for infants and children, the major source of exposure to pesticides is through diet. A study published in 2006 by Lu et al. measured the levels of organophosphorus pesticide exposure in 23 school children before and after replacing their diet with organic food. In this study it was found that levels of organophosphorus pesticide exposure dropped from negligible levels to undetectable levels when the children switched to an organic diet, the authors presented this reduction as a significant reduction in risk. The conclusions presented in Lu et al. were criticized in the literature as a case of bad scientific communication.

More specifically, claims related to pesticide residue of increased risk of infertility or lower sperm counts have not been supported by the evidence in the medical literature. Likewise the American Cancer Society (ACS) has stated their official position that "whether organic foods carry a lower risk of cancer because they are less likely to be contaminated by compounds that might cause cancer is largely unknown." Reviews have noted that the risks from microbiological sources or natural toxins are likely to be much more significant than short term or chronic risks from pesticide residues.

Microbiological Contamination

In looking at possible increased risk to safety from organic food consumption, reviews have found that although there may be increased risk from microbiological contamination due to increased manure use as fertilizer from organisms like *E. coli* O157:H7 during organic produce production, there is little evidence of actual incidence of outbreaks which can be positively blamed on organic food production. The 2011 Germany *E. coli* O104:H4 outbreak was blamed on organic farming of bean sprouts.

Economics

Demand for organic foods is primarily driven by concerns for personal health and for the environment. Global sales for organic foods climbed by more than 170 percent since 2002 reaching more than $63 billion in 2011 while certified organic farmland remained relatively small at less than 2 percent of total farmland under production, increasing in OECD and EU countries (which account for the majority of organic production) by 35 percent for the same time period. Organic products typically cost 10 to 40% more than similar conventionally produced products, to several times the price. Processed organic foods vary in price when compared to their conventional counterparts.

While organic food accounts for 1–2% of total food production worldwide, the organic food sales market is growing rapidly with between 5 and 10 percent of the food market share in the United States according to the Organic Trade Association, significantly outpacing sales growth volume in dollars of conventional food products. World organic food sales jumped from US $23 billion in 2002 to $63 billion in 2011.

Asia

Production and consumption of organic products is rising rapidly in Asia, and both China and India are becoming global producers of organic crops and a number of countries, particularly China and Japan, also becoming large consumers of organic food and drink. The disparity between production and demand, is leading to a two-tier organic food industry, typified by significant and growing imports of primary organic products such as dairy and beef from Australia, Europe, New Zealand and the United States.

China

- China's domestic organic market is the fourth largest in the world. The Chinese Organic Food Development Center estimated domestic sales of organic food products to be around US$500 million per annum as of 2013. This is predicted to increase by 30 percent to 50 percent in 2014. As of 2015, organic foods made up about 1% of the total Chinese food market.

- China is the world's biggest infant formula market with $12.4 billion in sales annually; of this, organic infant formula and baby food accounted for approximately 5.5 per cent of sales in 2011. Australian organic infant formula and baby food producer Bellamy's Organic have reported that their sales in this market grew 70 per cent annually over the period 2008-2013, while Organic Dairy Farmers of Australia, reported that exports of long-life organic milk to China had grown by 20 to 30 per cent per year over the same period.

Japan

- In 2010, the Japanese organic market was estimated to be around $1.3 billion.

North America

United States

As of October 2014, Trader Joe's is a market leader of organic grocery stores in the United States.

- In 2012 the total size of the organic food market in the United States was about $30 billion (out of the total market for organic and natural consumer products being about $81 billion)

- Organic food is the fastest growing sector of the American food industry.

- Organic food sales have grown by 17 to 20 percent a year in the early 2000s while sales of conventional food have grown only about 2 to 3 percent a year. The US organic market grew 9.5% in 2011, breaking the $30bn barrier for the first time, and continued to outpace sales of non-organic food.

- In 2003 organic products were available in nearly 20,000 natural food stores and 73% of conventional grocery stores.

- Organic products accounted for 3.7% of total food and beverage sales, and 11.4% of all fruit and vegetable sales in the year 2009.

- As of 2003, two thirds of organic milk and cream and half of organic cheese and yogurt are sold through conventional supermarkets.

- As of 2012, most independent organic food processors in the USA had been acquired by multinational firms.

- In order for a product to become USDA organic certified, the farmer cannot plant genetically modified seeds and livestock cannot eat geneti-

cally modified plants. Farmers must provide substantial evidence showing there was no genetic modification involved in the operation.

Canada

- Organic food sales surpassed $1 billion in 2006, accounting for 0.9% of
  food sales in Canada. By 2012, Canadian organic food sales reached $3
  billion.

- Organic food sales by grocery stores were 28% higher in 2006 than in 2005.

- British Columbians account for 13% of the Canadian population, but
  purchased 26% of the organic food sold in Canada in 2006.

Europe

Denmark

- In 2012, organic products accounted for 7.8% of the total retail consumption market in Denmark, the highest national market share in the
  world. Many public institutions have voluntarily committed themselves
  to buy some organic food and in Copenhagen 75 % of all food served in
  public institutions is organic. A governmental action plan initiated in
  2012-2014 aims at 60 % organic food in all public institutions across the
  country before 2020.

- In 1987, the first Danish Action Plan was implemented which was meant
  to support and stimulate farmers to switch from conventional food production systems to organic ones . Since then Denmark has constantly
  worked on further developing the market by promoting organic food
  and keeping prices low in comparison to conventional food products by
  offering farmers subvention and extra support if they choose to produce
  organic food. Then and even today is the bench mark for organic food
  policy and certification of organic food in the whole world. The new European Organic food label and organic food policy was developed based
  on the 1987 Danish Model.

Austria

- In 2011, 7.4% of all food products sold in Austrian supermarkets (including discount stores) were organic. In 2007, 8,000 different organic
  products were available.

Italy

- Since 2000, the use of some organic food is compulsory in Italian
  schools and hospitals. A 2002 law of the Emilia Romagna region imple-

mented in 2005, explicitly requires that the food in nursery and primary schools (from 3 months to 10 years) must be 100% organic, and the food in meals at schools, universities and hospitals must be at least 35% organic.

Poland

- In 2005 7 percent of Polish consumers buy food that was produced according to the EU-Eco-regulation. The value of the organic market is estimated at 50 million euros (2006).

Romania

- 70%–80% of the local organic production, amounting to 100 million euros in 2010, is exported. The organic products market grew to 50 million euros in 2010.

Switzerland

- As of 2012, 11 per cent of Swiss farms are organic. Bio Suisse, the Swiss organic producers' association, provides guidelines for organic farmers.

Ukraine

- In 2009 Ukraine was in 21st place in the world by area under cultivation of organic food. Much of its production of organic food is exported and not enough organic food is available on the national market to satisfy the rapidly increasing demand. The size of the internal market demand for organic products in Ukraine was estimated at over 5 billion euros in 2011, with rapid growth projected for this segment in the future. Multiple surveys show that the majority of the population of Ukraine is willing to pay more to buy organic food. On the other hand, many Ukrainians have traditionally maintained their own garden plots, and this may result in underestimation of how much organically produced food is actually consumed in Ukraine.

- The Law on Organic Production was passed by Ukraine's parliament in April 2011, which in addition to traditional demands for certified organic food also banned the use of GMOs or any products containing GMOs. However, the law was not signed by the President of Ukraine and in September 2011 it was repealed by the Verkhovna Rada itself. Attempts to pass a new law on organic food production took place throughout 2012.

United Kingdom

- Organic food sales increased from just over £100 million in 1993/94 to £1.21 billion in 2004 (an 11% increase on 2003). In 2010, the UK sales of organic products fell 5.9% to £1.73 billion. 86% of households buy organic products, the most popular categories being dairies (30.5% of sales) and fresh fruits and vegetables (23.2% of sales). 4.2% of UK farmland is organically managed.

Latin America

Cuba

- After the collapse of the Soviet Union in 1991, agricultural inputs that had previously been purchased from Eastern bloc countries were no longer available in Cuba, and many Cuban farms converted to organic methods out of necessity. Consequently, organic agriculture is a mainstream practice in Cuba, while it remains an alternative practice in most other countries. Although some products called organic in Cuba would not satisfy certification requirements in other countries (crops may be genetically modified, for example), Cuba exports organic citrus and citrus juices to EU markets that meet EU organic standards. Cuba's forced conversion to organic methods may position the country to be a global supplier of organic products.

Conservation Agriculture

Conservation agriculture (CA) can be defined by a statement given by the Food and Agricultural Organization of the United Nations as "a concept for resource-saving agricultural crop production that strives to achieve acceptable profits together with high and sustained production levels while concurrently conserving the environment" (FAO 2007).

Agriculture according to the New Standard Encyclopedia is "one of the most important sectors in the economies of most nations" (New Standard 1992). At the same time conservation is the use of resources in a manner that safely maintains a resource that can be used by humans. Conservation has become critical because the global population has increased over the years and more food needs to be produced every year (New Standard 1992). Sometimes referred to as "agricultural environmental management", conservation agriculture may be sanctioned and funded through conservation programs promulgated through agricultural legislation, such as the U.S. Farm Bill.

Key Principles

The Food and Agricultural Organization of the United Nations (FAO) has determined that CA has three key principles that producers (farmers) can proceed through in the process of CA. These three principles outline what conservationists and producers believe can be done to conserve what we use for a longer period of time.

The first key principle in CA is practicing minimum mechanical soil disturbance which is essential to maintaining minerals within the soil, stopping erosion, and preventing water loss from occurring within the soil. In the past agriculture has looked at soil tillage as a main process in the introduction of new crops to an area. It was believed that tilling the soil would increase fertility within the soil through mineralization that takes place in the soil. Also tilling of soil can cause severe erosion and crusting which leads to a decrease in soil fertility. Today tillage is seen as destroying organic matter that can be found within the soil cover. No-till farming has caught on as a process that can save soil organic levels for a longer period and still allow the soil to be productive for longer periods (FAO 2007). Additionally, the process of tilling can increase time and labor for producing that crop.

When no-till practices are followed, the producer sees a reduction in production cost for a certain crop. Tillage of the ground requires more money in order to fuel tractors or to provide feed for the animals pulling the plough. The producer sees a reduction in labor because he or she does not have to be in the fields as long as a conventional farmer.

The second key principle in CA is much like the first in dealing with protecting the soil. The principle of managing the top soil to create a permanent organic soil cover can allow for growth of organisms within the soil structure. This growth will break down the mulch that is left on the soil surface. The breaking down of this mulch will produce a high organic matter level which will act as a fertilizer for the soil surface. If CA practices were used done for many years and enough organic matter was being built up at the surface, then a layer of mulch would start to form. This layer helps prevent soil erosion from taking place and ruining the soil's profile or layout.

According to the article "The role of conservation agriculture and sustainable agriculture", the layer of mulch that is built up over time will become like a buffer zone between soil and mulch and this will help reduce wind and water erosion. With this comes the protection of the soil's surface when rain falls on the ground. Land that is not protected by a layer of mulch is left open to the elements (Hobbs et al. 2007). This type of ground cover also helps keep the temperature and moisture levels of the soil at a higher level rather than if it was tilled every year (FAO 2007).

The third principle is the practice of crop rotation with more than two species. According to an article published in the *Physiological Transactions of the Royal Society* called "The role of conservation agriculture and sustainable agriculture," crop rotation can be used best as a disease control against other preferred crops (Hobbs et al. 2007). This process will not allow pests such as insects and weeds to be set into a rotation with specific crops. Rotational crops will act as a natural insecticide and herbicide against specific crops. Not allowing insects or weeds to establish a pattern will help to eliminate problems with yield reduction and infestations within fields (FAO 2007). Crop rotation can also help build up soil infrastructure. Establishing crops in a rotation allows for an extensive buildup of rooting zones which will allow for better water infiltration (Hobbs et al. 2007).

Organic molecules in the soil break down into phosphates, nitrates and other beneficial elements which are thus better absorbed by plants. Plowing increases the amount of oxygen in the soil and increases the aerobic processes, hastening the breakdown of organic material. Thus more nutrients are available for the next crop but, at the same time, the soil is depleted more quickly of its nutrient reserves.

Examples

In conservation agriculture there are many examples that can be looked towards as a way of farming and at the same time conserving. These practices are well known by most producers. The process of no-till is one that follows the first principle of CA, causing minimal mechanical soil disturbance. No-till also brings other benefits to the producer . According to the FAO, tillage is one of the most "energy consuming" processes that can be used: It requires a lot of labor, time, and fuel to till. Producers can save 30% to 40% of time and labor by practicing the no-till process. (FAO 3020)

Besides conserving the soil, there are other examples of how CA is used. According to an article in *Science* called "Farming and the Fate of Wild Nature" there are two more kinds of CA . The practice of wildlife-friendly farming and land sparing are ideas for producers who are looking to practice better conservation towards biodiversity (Green, et al. 2005).

Wildlife-friendly Farming

Wildlife-friendly farming is a practice of setting aside land that will not be developed by the producer (farmer). This land will be set aside so that biodiversity has a chance to establish itself in areas with agricultural fields. At the same time, the producer is attempting to lower the amount of fertilizer and pesticides used on the fields so that organisms and microbial activity have a chance to establish themselves in the soil and habitat (Green, et al. 2005). But as in all systems, not all can be perfect. To create a habitat suitable for biodiversity something has to be reduced, and as in this case for agriculture farmers, yields can be reduced. This is where the second idea of land sparing can be looked on as an alternative manner

Land Sparing

Land sparing is another way that producer and conservationist can be on the same page. Land sparing advocates for the land that is being used for agricultural purposes to continue to produce crops at increased yield. With an increase in yield on all land that is in use, other land can be set aside for conservation and production for biodiversity. Agricultural land stays in production but would have to increase its yield potential to keep up with demand. Land that is not being put into agriculture would be used for conserving biodiversity (Green, et al. 2005).

Benefits

In the field of CA there are many benefits that both the producer and conservationist can obtain.

On the side of the conservationist, CA can be seen as beneficial because there is an effort to conserve what people use every day. Since agriculture is one of the most destructive forces against biodiversity, CA can change the way humans produce food and energy. With conservation come environmental benefits of CA. These benefits include less erosion possibilities, better water conservation, improvement in air quality due to lower emissions being produced, and a chance for larger biodiversity in a given area.

On the side of the producer and/or farmer, CA can eventually do all that is done in conventional agriculture, and it can conserve better than conventional agriculture. CA according to Theodor Friedrich, who is a specialist in CA, believes "Farmers like it because it gives them a means of conserving, improving, and making more efficient use of their natural resources" (FAO 2006). Producers will find that the benefits of CA will come later rather than sooner. Since CA takes time to build up enough organic matter and have soils become their own fertilizer, the process does not start to work overnight. But if producers make it through the first few years of production, results will start to become more satisfactory.

CA is shown to have even higher yields and higher outputs than conventional agriculture once it has been established over long periods. Also, a producer has the benefit of knowing that the soil in which his crops are grown is a renewable resource. According to New Standard Encyclopedia, soils are a renewable resource, which means that whatever is taken out of the soil can be put back over time (New Standard 1992). As long as good soil upkeep is maintained, the soil will continue to renew itself. This could be very beneficial to a producer who is practicing CA and is looking to keep soils at a productive level for an extended time.

The farmer and/or producer can use this same land in another way when crops have been harvested. The introduction of grazing livestock to a field that once held crops can be beneficial for the producer and also the field itself. Livestock can be used as a natural fertilizer for a producer's field which will then be beneficial for the producer the next year when crops are planted once again. The practice of grazing livestock using CA helps the farmer who raises crops on that field and the farmer who raises the livestock that graze off that field. Livestock produce compost or manure which are a great help in generating soil fertility (Pawley W.H. 1963). The practices of CA and grazing livestock on a field for many years can allow for better yields in the following years as long as these practices continue to be followed.

The FAO believes that there are three major benefits from CA:

- Within fields that are controlled by CA the producer will see an increase in organic matter.

- Increase in water conservation due to the layer of organic matter and ground cover to help eliminate transportation and access runoff.

- Improvement of soil structure and rooting zone.

Future Development

As in any other business, producers and conservationists are always looking towards the future. In this case CA is a very important process to be looked at for future generation. There are many organizations that have been created to help educate and inform producers and conservationists in the world of CA. These organizations can help to inform, conduct research, and buy land in order to preserve animals and plants (New Standard 1992).

Another way in which CA is looking to the future is through prevention. According to the *European Journal of Agronomy* producers are looking for ways to reduce leaching problems within their fields. These producers are using the same principles within CA, in that they are leaving cover over their fields in order to save fields from erosion and leaching of chemicals (Kirchmann & Thorvaldsson 2000). Processes and studies like this are allowing for a better understanding of how to conserve what we are using and finding ways to put back something that may have been lost before.

In the same journal article is presented another way in which producers and conservationists are looking towards the future. Circulation of plant nutrients can be a vital part for conserving the future. An example of this would be the use of animal manure. This process has been used for quite some time now, but the future is looking towards ways to handle and conserve nutrients within manure for a longer time. But besides animal waste, food and urban waste are also being looked towards as a way to use growth within CA (Kirchmann & Thorvaldsson 2000). Turning these products from waste to being used to grow crops and improve yields is something that would be beneficial for conservationists and producers.

Agri-environment Schemes

In 1992, 'agri-environment schemes' became compulsory for all European Union Member States. In the following years the main purpose of these schemes changed slightly. Initially, they sought to protect threatened habitats, but gradually shifted their focus to the prevention of the loss of wildlife from agricultural landscapes. Most recently, the schemes are placing more emphasis on improving the services that the land can provide to humans (e.g. pollination). Overall, farmers involved in the scheme aim to practice environmentally friendlier farming techniques such as: reducing the use of pesticides, managing or altering their land to increase more wildlife friendly habitats (e.g. increasing areas of trees and bushes), reducing irrigation, conserving soil, and organic farming. As the changes in practices that ensure the protection of the environment are costly to farmers, the EU developed agri-environment schemes to financially compensate individual farmers for applying these changes and therefore increased the implementation of conservation agriculture. The schemes are voluntary for farmers. Once joined, they commit to a minimum of five years during which they have to adopt various sustainable farming techniques. According to the Euro-stat website, in 2009 the agricultural area enrolled in agri-environment schemes covered 38.5 million hectares (20.9% of agricultural land in the 27 member states of the EU at the time) (Agri-environmental indicator 2015). The European Commission spent a total of €3.23 billion on agri-environment schemes in 2012, significantly exceeding the cost of managing special sites of conservation (Natura 2000) that year, which came to a total of €39.6 million (Batáry et al. 2015). There are two main types of agri-environment schemes which have shown different outcomes. Out-of-production schemes tend to be used in extensive farming practices (where the farming land is widespread and less intensive farming is practiced), and focus on improving or setting land aside that will not be used for the production of food, for example, the addition of wildflower strips. In-production schemes (used for a smaller scale, but more intensively farmed land) focus on the sustainable management of arable crops or grassland, for example reduction of pesticides, reduction of grassland mowing, and most commonly, organic farming. In a 2015 review of studies examining the effects of the two schemes, it was found that out-of-production schemes had a higher success rate at enhancing the number of thriving species around the land. The reason behind this is thought to be the scheme's focus on enhancing specific species by providing them with more unaltered habitats, which results in more food resources for the specific species. On the other hand, in-production schemes attempt to enhance the quality of the land in general, and are thus less species specific. Based on the findings, the reviewers suggest that schemes which more specifically target the declining groups of species, may be more effective. The findings and the targets will be implemented between 2015 and 2020, so that by 2025, the effectiveness of these schemes can be re-assessed and will have increased significantly (Batáry et al. 2015).

Problems

As much as conservation agriculture can benefit the world, there are some problems that come with it. There are many reasons why conservation agriculture cannot always be a win-win situation.

There are not enough people who can financially turn from conventional farming to conservation. The process of CA takes time; when a producer first becomes a conservationist, the results can be a financial loss to them. CA is based upon establishing an organic layer and producing its own fertilizer and this may take time. It can be many years before a producer will start to see better yields than he/she has had previously. Another financial undertaking is purchasing of new equipment. When starting to use CA, a producer may have to buy new planters or drills in order to produce effectively. These financial tasks are ones that may impact whether or not a producer decides to switch to CA or not. It also that many of the farmers have to cut the tree to farming there because some places are not distant to being farming. So, it also effect the environment. Government should provide the places to the farmers for farming.

With the struggle to adapt comes the struggle to make CA grow across the globe. CA has not spread as quickly as most conservationists would like. The reason for this is because there is not enough pressure for producers in places such as North America to change their way of living to a more conservationist outlook. But in the tropics there is more pressure to change to conservation areas because of the limited resources that are available. Places like Europe have also started to catch onto the ideas and principles of CA, but still nothing much is being done to change due to there being a minimal amount of pressure for people to change their ways of living (FAO 2006).

With CA comes the idea of producing enough food. With cutting back in fertilizer, not tilling the ground, and other processes comes the responsibility to feed the world. According to the Population Reference Bureau, there were around 6.08 billion people on Earth in the year 2000. By 2050 there will be an estimated 9.1 billion people. With this increase comes the responsibility for producers to increase food supply using the same or less land than we use today. Problems arise in the fact that if CA farms do not produce as much as conventional farms, this leaves the world with less food for more people.

No-till Farming

Young soybean plants thrive in and are protected by the residue of a wheat crop. This form of no till farming provides good protection for the soil from erosion and helps retain moisture for the new crop.

No-till farming (also called zero tillage or direct drilling) is a way of growing crops or pasture from year to year without disturbing the soil through tillage. No-till is an agricultural technique which increases the amount of water that infiltrates into the soil and increases organic matter retention and cycling of nutrients in the soil. In many agricultural regions it can reduce or eliminate soil erosion. It increases the amount and variety of life in and on the soil, including disease-causing organisms and disease suppression organisms. The most powerful benefit of no-tillage is improvement in soil biological fertility, making soils more resilient. Farm operations are made much more efficient, particularly improved time of sowing and better trafficability of farm operations.

Background

Tilling is the process of removing plants or plant debris, usually for the purposes of planting more desirable species. This tilling can result in a flat seed bed or one that has formed areas, such as rows or raised beds, to enhance the growth of desired plants. It is an ancient technique with clear evidence of its use since at least 3000 B.C.

The effects of tillage can include soil compaction; loss of organic matter; degradation of soil aggregates; death or disruption of soil microbes and other organisms including mycorrhizae, arthropods, and earthworms; and soil erosion where topsoil is washed or blown away.

Origin of No-till for Modern Farms

The idea of modern no-till farming started in the 1940s with Edward H. Faulkner, author of *Plowman's Folly*, but it wasn't until the development of several chemicals after WWII that various researchers and farmers started to try out the idea. The first adopters of no-till include Klingman (North Carolina), Edward Faulkner, L.A. Porter (New Zealand), Harry and Lawrence Young (Herndon, Kentucky), the Instituto de Pesquisas Agropecuarias Meridional (1971 in Brazil) with Herbert Bartz.

Adoption Rate in the United States

No-till farming is widely used in the United States and the number of acres managed in this way continues to grow. This growth is supported by a decrease in costs related to tillage; no-till management results in fewer passes with equipment for approximately equal harvests, and the crop residue prevents evaporation of rainfall and increases water infiltration into the soil.

Issues

Profit, Economics, Yield

Studies have found that no-till farming can be more profitable if performed correctly.

Less tillage of the soil reduces labour, fuel, irrigation and machinery costs. No-till can increase yield because of higher water infiltration and storage capacity, and less erosion. Another benefit of no-till is that because of the higher water content, instead of leaving a field fallow it can make economic sense to plant another crop instead.

As sustainable agriculture becomes more popular, monetary grants and awards are becoming readily available to farmers who practice conservation tillage. Some large energy corporations which are among the greatest generators of fossil-fuel-related pollution may purchase carbon credits, which can encourage farmers to engage in conservation tillage. Under such schemes, the farmers' land is legally redefined as a carbon sink for the power generators' emissions. This helps the farmer in several ways, and it helps the energy companies meet regulatory demands for reduction of pollution, specifically carbon emissions.

No-till farming can increase organic (carbon based) matter in the soil, which is a form of carbon sequestration. However, there is debate over whether this increased sequestration detected in scientific studies of no-till agriculture is actually occurring, or is due to flawed testing methods or other factors. Regardless of this debate, a case can still be made for no-till, in the form of reduction in fossil fuel use, less erosion and better soil quality.

Environmental

Carbon (Air and Soil) and other Greenhouse Gases

No-till farming has carbon sequestration potential through storage of soil organic matter in the soil of crop fields. Whereas, when soil is tilled by machinery, the soil layers invert, air mixes in, and soil microbial activity dramatically increases over baseline levels. Tilling results in soil organic matter being broken down much more rapidly, and carbon is lost from the soil into the atmosphere. In addition to the effect on soil from tilling, emissions from the farm tractors increases carbon dioxide levels in the atmosphere.

Cropland soils are ideal for use as a carbon sink, since they have been depleted of carbon in most areas. It is estimated that 78 billion metric tonnes of carbon that was trapped in the soil has been released because of tillage. Conventional farming practices that rely on tillage have removed carbon from the soil ecosystem by removing crop residues such as left over corn stalks, and through the addition of chemical fertilizers which have the above-mentioned effects on soil microbes. By eliminating tillage, crop residues decompose where they lie, and by growing winter cover crops, carbon loss can be slowed and eventually reversed.

Nonetheless, a growing body of research is showing that no-till systems lose carbon stocks over time. Regarding a 2014 study of which he was principal investigator, University of Illinois soil scientist Ken Olson said this differing result occurs in part because tested soil samples need to include the full depth of rooting; 1–2 meters deep. He said, "That no-till subsurface layer is often losing more soil organic carbon stock over time than is gained in the surface layer." Also, there has not been a uniform defi-nition of soil organic carbon sequestration among researchers. The study concludes, "Additional investments in SOC research is needed to better understand the agricultur-al management practices that are most likely to sequester SOC or at least retain more net SOC stocks."

In addition to keeping carbon in the soil, no-till farming reduces nitrous oxide (N_2O) emissions by 40-70%, depending on rotation. Nitrous oxide is a potent greenhouse gas, 300 times stronger than CO2, and stays in the atmosphere for 120 years. Fertilizing farmlands with (excessive) nitrogen increases the release of nitrous oxide.

Soil and Water

No-till farming improves soil quality (soil function), carbon, organic matter, aggregates, protecting the soil from erosion, evaporation of water, and structural breakdown. A re-duction in tillage passes helps prevent the compaction of soil.

Recently, researchers at the Agricultural Research Service of the United States De-partment of Agriculture found that no-till farming makes soil much more stable than plowed soil. Their conclusions draw from over 19 years of collaborated tillage studies. No-till stores more carbon in the soil and carbon in the form of organic matter is a key factor in holding soil particles together. The first inch of no-till soil is two to seven times less vulnerable than that of plowed soil. The practice of no-till farming is especially beneficial to Great Plains farmers because of its avoidance of erosion.

Crop residues left intact help both natural precipitation and irrigation water infiltrate the soil where it can be used. The crop residue left on the soil surface also limits evap-oration, conserving water for plant growth. Soil compaction and no tillage-pan, soil absorbs more water and plants are able to grow their roots deeper into the soil and suck up more water.

Tilling a field reduces the amount of water, via evaporation, around 1/3 to 3/4 inches (0.85 to 1.9 cm) per pass. By no-tilling, this water stays in the soil, available to the plants.

Soil Biota, Wildlife, etc.

In no-till farming the soil is left intact and crop residue is left on the field. Therefore, soil layers, and in turn soil biota, are conserved in their natural state. No-tilled fields of-ten have more beneficial insects and annelids, a higher microbial content, and a greater amount of soil organic material. Since there is no ploughing there is less airborne dust.

No-till farming increases the amount and variety of wildlife. This is the result of im-proved cover, reduced traffic and the reduced chance of destroying ground nesting birds and animals (plowing destroys all of them).

Albedo

Tillage lowers the albedo of croplands. The potential for global cooling as a result of increased Albedo in no till croplands is similar in magnitude to the biogeochemical (carbon sequestration) potential.

Historical Artifacts

Tilling regularly damages ancient structures under the soil such as long barrows. In the UK, half of the long barrows in Gloucestershire and almost all the burial mounds in Essex have been damaged. According to English Heritage modern tillage techniques have done as much damage in the last six decades as traditional tilling did in the previous six centuries. By using no-till methods these structures can be preserved and can be properly investigated instead of being destroyed.

Cost

Equipment

No-till farming requires specialized seeding equipment designed to plant seeds into undisturbed crop residues and soil. If the farmer has equipment designed for tillage farming, purchasing new equipment (seed drills for example) would be expensive and while the cost could be offset by selling off plows, etc. doing so is not usually done until the farmer decides to switch completely over (after trying it out for a few years). This would result in more money being invested into equipment in the short term (until old equipment is sold off).

Drainage

If a soil has poor drainage, it may need drainage tiles or other devices in order to help with the removal of excess water under no-till. Farmers should remember that water infiltration will improve after several years of a field being in no-till farming, so they may want to wait until 5–8 years have passed to see if the problems persists before deciding to invest in such an expensive system.

Gullies

Gullies can be a problem in the long-term. While much less soil is displaced by using no-till farming, any drainage gulleys that do form will get deeper each year since they aren't being smoothed out by plowing. This may necessitate either sod drainways, waterways, permanent drainways, cover crops, etc. Gully formation can be avoided entirely with proper water management practices, including the creation of swales on contour.

Increased Chemical Use

One of the purposes of tilling is to remove weeds. No-till farming does change weed composition drastically. Faster growing weeds may no longer be a problem in the face of increased competition, but shrubs and trees may begin to grow eventually.

Some farmers attack this problem with a "burn-down" herbicide such as glyphosate in lieu of tillage for seedbed preparation and because of this, no-till is often associated with increased chemical use in comparison to traditional tillage based methods of crop production. However, there are many agroecological alternatives to increased chemical use, such as winter cover crops and the mulch cover they provide, soil solarization or burning.

Management

No-till farming requires some different skills to achieve success. As with any production system, if no-till isn't done correctly, yields can drop. A combination of technique, equipment, pesticides, crop rotation, fertilization, and irrigation have to be used for local conditions.

Cover Crops

Cover crops are used occasionally in no-till farming to help control weeds and increase nutrients in the soil (by using legumes) or by using plants with long roots to pull mobile nutrients back up to the surface from lower layers of the soil. Farmers experimenting with organic no-till use cover crops instead of tillage for controlling weeds, and are developing various methods to kill the cover crops (rollers, crimper, choppers, etc.) so that the newly planted crops can get enough light, water, nutrients, etc.

Disease, Pathogens, Insects and the use of Crop Rotations

With no-till farming, residue from the previous years crops lie on the surface of the field, cooling it and increasing the moisture. This can cause increased or decreased or variations of diseases that occur, but not necessarily at a higher or lower rate than conventional tillage. In order to help eliminate weed, pest and disease problems, crop rotations are used. By rotating the crops on a multi-year cycle, pests and diseases will decrease since the pests will no longer have a food supply to support their numbers.

Organic No-till Technique: The Cardboard Method

Some farmers who prefer to pursue a chemical-free management practice often rely on the use of normal, non-dyed corrugated cardboard for use on seed-beds and vegetable areas. Used correctly, cardboard placed on a specific area can

1. keep important fungal hyphae and microorganisms in the soil intact

2. prevent recurring weeds from popping up

3. increase residual nitrogen and plant nutrients by top-composting plant residues and

4. create valuable topsoil that is well suited for next years seeds or transplants.

The plant residues (left over plant matter originating from cover crops, grass clippings, original plant life etc.) will rot while underneath the cardboard so long as it remains sufficiently moist. This rotting attracts worms and other beneficial microorganisms to the site of decomposition, and over a series of a few seasons (usually Spring-->Fall or Fall-->Spring) and up to a few years, will create a layer of rich topsoil. Plants can then be direct seeded into the soil come spring, or holes can be cut into the cardboard to allow for transplantation. Using this method in conjunction with other sustainable practices such as composting/vermicompost, cover crops and rotations are often considered beneficial to both land and those who take from it.

Water Issues

No-till farming dramatically reduces the amount of erosion in a field. While much less soil is displaced, any gullies that do form will get deeper each year instead of being smoothed out by regular plowing. This may necessitate either sod drainways, waterways, permanent drainways, cover crops, etc.

A problem that occurs in some fields is water saturation in soils. Switching to no-till farming will correct the drainage the because of the qualities of soil under continuous no-till include a higher water infiltration rate.

Equipment

It is very important to have planting equipment that can properly penetrate through the residue, into the soil and prepare a good seedbed. Switching to no-till reduces the maximum amount of power needed from farm tractors, which means that a farmer can farm under no-till with a smaller tractor than under tilling. Using a smaller, lighter tractor has the added benefit of reducing compaction.

Soil Temperature

Another problem that growers face is that in the spring the soil will take longer to warm and dry, which may delay planting to a less ideal future date. One reason why the soil is slower to warm is that the field absorbs less solar energy as the residue covering the soil is a much lighter color than the black soil which would be exposed in conventional tillage. This can be managed by using row cleaners on a planter. Since the soil can be cooler, harvest can occur a few days later than a conventionally tilled field. Note: A cooler soil is also a benefit because water doesn't evaporate as fast.

Residue

On some crops, like continuous no-till corn, the thickness of the residue on the surface of the field can become a problem without proper preparation and/or equipment.

Fertilizer

One of the most common yield reducers is nitrogen being immobilized in the crop residue, which can take a few months to several years to decompose, depending on the crop's C to N ratio and the local environment. Fertilizer needs to be applied at a higher rate during the transition period while the soil rebuilds its organic matter. The nutrients in the organic matter will be eventually released back into the soil, so this is only a concern during the transition time frame (4–5 years for Kansas, USA). An innovative solution to this problem is to integrate animal husbandry in various ways to aid in the decomposition cycle.

Misconceptions

Need to Fluff the Soil

Although no-till farming often causes a slight increase in soil bulk density, periodic tilling is not needed to "fluff" the soil back up. No-till farming mimics the natural conditions under which most soils formed more closely than any other method of farming, in that the soil is left undisturbed except to place seeds in a position to germinate.

Similar Terms

No-till farming is not equivalent to conservation tillage or strip tillage. Conservation tillage is a group of practices that reduce the amount of tillage needed. No-till and strip tillage are both forms of conservation tillage. No-till is the practice of never tilling a field. Tilling every other year is called rotational tillage.

Shifting Cultivation

Shifting cultivation is an agricultural system in which plots of land are cultivated temporarily, then abandoned and allowed to revert to their natural vegetation while the cultivator moves on to another plot. The period of cultivation is usually terminated when the soil shows signs of exhaustion or, more commonly, when the field is overrun by weeds. The length of time that a field is cultivated is usually shorter than the period over which the land is allowed to regenerate by lying fallow. This technique is often used in LEDCs (Less Economically Developed Countries) or LICs (Low Income Countries). In some areas, cultivators use a practice of slash-and-burn as one element of their farming cycle. Others employ land clearing without any burning, and some cultivators are purely migratory and do not use any cyclical method on a given plot. Sometimes no slashing at all is needed where regrowth is purely of grasses, an outcome not uncommon when soils are near exhaustion and need to lie fallow. In shifting agriculture, after two or three years of producing vegetable and grain crops on cleared land, the migrants abandon it for another plot. Trees and bushes are cleared by slashing, and the remaining vegetation is burnt. The ashes add potash to the soil. Then the seeds are sown after the rains.

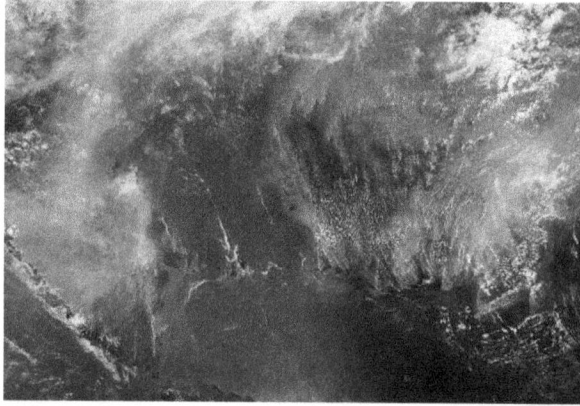

Slash-and-burn based shifting cultivation is a widespread historical practice in southeast Asia. Above is a satellite image of Sumatra and Borneo showing shift cultivation fires from October 2006.

Advantages of Slash-and-burn Method

Slash-and-burn is a very sustainable technique. It differs a lot from commercial farming, because once the trees are burned, there is very fertile fine ash that deposits along the humus, meaning that by the time the other fields are burned, the soil has time to reassemble nutriments, in order to make cultural activity possible. Although slash-and-burn is a very useful technique, there are other ways of fertilizing soil. By planting beans, the soil will regenerate much faster, due to the production of nitrogen in their roots.

Political Ecology of Shifting Cultivation

Shifting cultivation is a form of agriculture or a cultivation system, in which, at any particular point in time, a minority of 'fields' are in cultivation and a majority are in various stages of natural re-growth. Over time, fields are cultivated for a relatively short time, and allowed to recover, or are fallowed, for a relatively long time. Eventually a previously cultivated field will be cleared of the natural vegetation and planted in crops again. Fields in established and stable shifting cultivation systems are cultivated and fallowed cyclically.This type of farming is called jhumming in India.

Fallow fields are not unproductive. During the fallow period, shifting cultivators use the successive vegetation species widely for timber for fencing and construction, firewood, thatching, ropes, clothing, tools, carrying devices and medicines. It is common for fruit and nut trees to be planted in fallow fields to the extent that parts of some fallows are in fact orchards. Soil-enhancing shrub or tree species may be planted or protected from slashing or burning in fallows. Many of these species have been shown to fix nitrogen. Fallows commonly contain plants that attract birds and animals and are important for hunting. But perhaps most importantly, tree fallows protect soil against physical erosion and draw nutrients to the surface from deep in the soil profile.

The relationship between the time the land is cultivated and the time it is fallowed are critical to the stability of shifting cultivation systems. These parameters determine whether or not the shifting cultivation system as a whole suffers a net loss of nutrients over time. A system in which there is a net loss of nutrients with each cycle will eventually lead to a degradation of resources unless actions are taken to arrest the losses. In some cases soil can be irreversibly exhausted (including erosion as well as nutrient loss) in less than a decade.

The longer a field is cropped, the greater the loss of soil organic matter, cation-exchange-capacity and in nitrogen and phosphorus, the greater the increase in acidity, the more likely soil porosity and infiltration capacity is reduced and the greater the loss of seeds of naturally occurring plant species from soil seed banks. In a stable shifting cultivation system, the fallow is long enough for the natural vegetation to recover to the state that it was in before it was cleared, and for the soil to recover to the condition it was in before cropping began. During fallow periods soil temperatures are lower, wind and water erosion is much reduced, nutrient cycling becomes closed again, nutrients are extracted from the subsoil, soil fauna decreases, acidity is reduced, soil structure, texture and moisture characteristics improve and seed banks are replenished.

The secondary forests created by shifting cultivation are commonly richer in plant and animal resources useful to humans than primary forests, even though they are much less bio-diverse. Shifting cultivators view the forest as an agricultural landscape of fields at various stages in a regular cycle. People unused to living in forests cannot see the fields for the trees. Rather they perceive an apparently chaotic landscape in which trees are cut and burned randomly and so they characterise shifting cultivation as ephemeral or 'pre-agricultural', as 'primitive' and as a stage to be progressed beyond. Shifting agriculture is none of these things. Stable shifting cultivation systems are highly variable, closely adapted to micro-environments and are carefully managed by farmers during both the cropping and fallow stages. Shifting cultivators may possess a highly developed knowledge and understanding of their local environments and of the crops and native plant species they exploit. Complex and highly adaptive land tenure systems sometimes exist under shifting cultivation. Introduced crops for food and as cash have been skillfully integrated into some shifting cultivation systems.

Shifting Cultivation in Europe

Shifting cultivation was still being practised as a viable and stable form of agriculture in many parts of Europe and east into Siberia at the end of the 19th century and in some places well into the 20th century. In the Ruhr in the late 1860s a forest-field rotation system known as *Reutbergwirtschaft* was using a 16-year cycle of clearing, cropping and fallowing with trees to produce bark for tanneries, wood for charcoal and rye for flour (Darby 1956, 200). Swidden farming was practised in Siberia at least until the 1930s, using specially selected varieties of "swidden-rye" (Steensberg 1993, 98). In Eastern Europe and Northern Russia the main swidden crops were turnips, barley, flax, rye, wheat, oats, radishes and millet. Cropping periods were usually one year, but were extended to two or three years on very favourable soils. Fallow periods were between 20 and 40 years (Linnard 1970, 195). In Finland in 1949, Steensberg (1993, 111) observed the clearing and burning of a 60,000 square metres (15 acres) swidden 440 km north of Helsinki. Birch and pine trees had been cleared over a period of a year and the logs sold for cash. A fallow of alder (Alnus) was encouraged to improve soil conditions. After the burn, turnip was sown for sale and for cattle feed. Shifting cultivation was disappearing in this part of Finland because of a loss of agricultural labour to the industries of the towns. Steensberg (1993, 110-152) provides eye-witness descriptions of shifting cultivation being practised in Sweden in the 20th century, and in Estonia, Poland, the Caucasus, Serbia, Bosnia, Hungary, Switzerland, Austria and Germany in the 1930s to the 1950s.

That these agricultural practices survived from the Neolithic into the middle of the 20th century amidst the sweeping changes that occurred in Europe over that period, suggests they were adaptive and in themselves, were not massively destructive of the environments in which they were practised. This raises the question: if shifting cultivation did not lead to the disappearance of European forests, what did?

The earliest written accounts of forest destruction in Southern Europe begin around 1000 BC in the histories of Homer, Thucydides and Plato and in Strabo's Geography. Forests were exploited for ship building, and urban development, the manufacture of casks, pitch and charcoal, as well as being cleared for agriculture. The intensification of trade and as a result of warfare, increased the demand for ships which were manufactured completely from forest products. Although goat herding is singled out as an important cause of environmental degradation, a more important cause of forest destruction was the practice in some places of granting ownership rights to those who clear felled forests and brought the land into permanent cultivation. Evidence that circumstances other than agriculture were the major causes for forest destruction was the recovery of tree cover in many parts of the Roman empire from 400 BC to around 500 AD following the collapse of Roman economy and industry. Darby observes that by 400 AD "land that had once been tilled became derelict and overgrown" and quotes Lactantius who wrote that in many places "cultivated land became forest" (Darby 1956, 186). The other major cause of forest destruction in the Mediterranean environment with its hot dry summers were wild fires that became more common following human interference in the forests.

In Central and Northern Europe the use of stone tools and fire in agriculture is well established in the palynological and archaeological record from the Neolithic. Here, just as in Southern Europe, the demands of more intensive agriculture and the invention of the plough, trading, mining and smelting, tanning, building and construction in the growing towns and constant warfare, including the demands of naval shipbuilding, were more important forces behind the destruction of the forests than was shifting cultivation.

By the Middle Ages in Europe, large areas of forest were being cleared and converted into arable land in association with the development of feudal tenurial practices. From the 16th to the 18th centuries, the demands of iron smelters for charcoal, increasing industrial developments and the discovery and expansion of colonial empires as well as incessant warfare that increased the demand for shipping to levels never previously reached, all combined to deforest Europe. With the loss of the forest, so shifting cultivation became restricted to the peripheral places of Europe, where permanent agriculture was uneconomic, transport costs constrained logging or terrain prevented the use of draught animals or tractors. It has disappeared from even these refuges since 1945, as agriculture has become increasingly capital intensive, rural areas have become depopulated and the remnant European forests themselves have been revalued economically and socially.

Simple Societies, Shifting Cultivation and Environmental Change

Shifting cultivation in Indonesia. A new crop is sprouting through the burnt soil.

A growing body of palynological evidence finds that simple human societies brought about extensive changes to their environments before the establishment of any sort of state, feudal or capitalist, and before the development of large scale mining, smelting or shipbuilding industries. In these societies agriculture was the driving force in the economy and shifting cultivation was the most common type of agriculture practiced. By examining the relationships between social and economic change and agricultural change in these societies, insights can be gained on contemporary social and economic change and global environment change, and the place of shifting cultivation in those relationship.

As early as 1930 questions about relationships between the rise and fall of the Mayan civilization of the Yucatán Peninsula and shifting cultivation were raised and continue to be debated today. Archaeological evidence suggests the development of Mayan society and economy began around 250 AD. A mere 700 years later it reached its apogee, by which time the population may have reached 2,000,000 people. There followed a precipitous decline that left the great cities and ceremonial centres vacant and overgrown with jungle vegetation. The causes of this decline are uncertain; but warfare and the exhaustion of agricultural land are commonly cited (Meggers 1954; Dumond 1961; Turner 1974). More recent work suggests the Maya may have, in suitable places, developed irrigation systems and more intensive agricultural practices (Humphries 1993).

Similar paths appear to have been followed by Polynesian settlers in New Zealand and the Pacific Islands, who within 500 years of their arrival around 1100 AD turned substantial areas from forest into scrub and fern and in the process caused the elimination of numerous species of birds and animals (Kirch and Hunt 1997). In the restricted environments of the Pacific islands, including Fiji and Hawaii, early extensive erosion and change of vegetation is presumed to have been caused by shifting cultivation on slopes. Soils washed from slopes were deposited in valley bottoms as a rich, swampy alluvium. These new environments were then exploited to develop intensive, irrigated fields. The change from shifting cultivation to intensive irrigated fields occurred in association with a rapid growth in population and the development of elaborate and highly stratified chiefdoms (Kirch 1984). In the larger, temperate latitude, islands of New Zealand the presumed course of events took a different path. There the stimulus for population growth was the hunting of large birds to extinction, during which time forests in drier areas were destroyed by burning, followed the development of intensive agriculture in favorable environments, based mainly on sweet potato (Ipomoea batatas) and a reliance on the gathering of two main wild plant species in less favorable environments. These changes, as in the smaller islands, were accompanied by population growth, the competition for the occupation of the best environments, complexity in social organization, and endemic warfare (Anderson 1997).

The record of humanly induced changes in environments is longer in New Guinea than in most places. Agricultural activities probably began 5,000 to 9,000 years ago. However, the most spectacular changes, in both societies and environments, are believed to have occurred in the central highlands of the island within the last 1,000 years, in association with the introduction of a crop new to New Guinea, the sweet potato (Golson 1982a; 1982b). One of the most striking signals of the relatively recent intensification of agriculture is the sudden increase in sedimentation rates in small lakes.

The root question posed by these and the numerous other examples that could be cited of simple societies that have intensified their agricultural systems in association with increases in population and social complexity is not whether or how shifting cultivation was responsible for the extensive changes to landscapes and environments. Rather it is why simple societies of shifting cultivators in the tropical forest of Yucatán, or the highlands of New Guinea, began to grow in numbers and to develop stratified and sometimes complex social hierarchies?

At first sight, the greatest stimulus to the intensification of a shifting cultivation system is a growth in population. If no other changes occur within the system, for each extra person to be fed from the system, a small extra amount of land must be cultivated. The total amount of land available is the land being presently cropped and all of the land in fallow. If the area occupied by the system is not expanded into previously unused land, then either the cropping period must be extended or the fallow period shortened.

At least two problems exist with the population growth hypothesis. First, population growth in most pre-industrial shifting cultivator societies has been shown to be very low over the long term. Second, no human societies are known where people work only to eat. People engage in social relations with each other and agricultural produce is used in the conduct of these relationships.

These relationships are the focus of two attempts to understand the nexus between human societies and their environments, one an explanation of a particular situation and the other a general exploration of the problem.

1. Feedback Loops

In a study of the Duna in the Southern Highlands of New Guinea, a group in the process of moving from shifting cultivation into permanent field agriculture post sweet potato, Modjeska (1982) argued for the development of two "self amplifying feed back loops" of ecological and social causation. The trigger to the changes was very slow population growth and the slow expansion of agriculture to meet the demands of this growth. This set in motion the first feedback loop, the "use-value" loop. As more forest was cleared there was a decline in wild food resources and protein produced from hunting, which was substituted for by an increase in domestic pig raising. An increase in domestic pigs required a further expansion in agriculture. The greater protein available from the larger number of pigs increased human fertility and survival rates and resulted in faster population growth.

The outcome of the operation of the two loops, one bringing about ecological change and the other social and economic change, is an expanding and intensifying agricultural system, the conversion of forest to grassland, a population growing at an increasing rate and expanding geographically and a society that is increasing in complexity and stratification.

2. Resources are Cultural Appraisals

The second attempt to explain the relationships between simple agricultural societies and their environments is that of Ellen (1982, 252-270). Ellen does not attempt to separate use-values from social production. He argues that almost all of the materials required by humans to live (with perhaps the exception of air) are obtained through social relations of production and that these relations proliferate and are modified in numerous ways. The values that humans attribute to items produced from the environment arise out of cultural arrangements and not from the objects themselves, a restatement of Carl Sauer's dictum that "resources are cultural appraisals". Humans frequently translate actual objects into culturally conceived forms, an example being the translation by the Duna of the pig into an item of compensation and redemption. As a result, two fundamental processes underlie the ecology of human social systems: First, the obtaining of materials from the environment and their alteration and circulation through social relations, and second, giving the material a value which will affect how important it is to obtain it, circulate it or alter it. Environmental pressures are thus mediated through social relations.

Transitions in ecological systems and in social systems do not proceed at the same rate. The rate of phylogenetic change is determined mainly by natural selection and partly by human interference and adaptation, such as for example, the domestication of a wild species. Humans however have the ability to learn and to communicate their knowledge to each other and across generations. If most social systems have the tendency to increase in complexity they will, sooner or later, come into conflict with, or into "contradiction" (Friedman 1979, 1982) with their environments. What happens around the point of "contradiction" will determine the extent of the environmental degradation that will occur. Of particular importance is the ability of the society to change, to invent or to innovate technologically and sociologically, in order to overcome the "contradiction" without incurring continuing environmental degradation, or social disintegration.

An economic study of what occurs at the points of conflict with specific reference to shifting cultivation is that of Esther Boserup (1965). Boserup argues that low intensity farming, extensive shifting cultivation for example, has lower labor costs than more intensive farming systems. This assertion remains controversial. She also argues that given a choice, a human group will always choose the technique which has the lowest absolute labor cost rather than the highest yield. But at the point of conflict, yields will have become unsatisfactory. Boserup argues, contra Malthus, that rather than population always overwhelming resources, that humans will invent a new agricultural technique or adopt an existing innovation that will boost yields and that is adapted to the new environmental conditions created by the degradation which has occurred already, even though they will pay for the increases in higher labor costs. Examples of such changes are the adoption of new higher yielding crops, the exchanging of a digging stick for a hoe, or a hoe for a plough, or the development of irrigation systems. The controversy over Boserup's proposal is in part over whether intensive systems are more costly in labor terms, and whether humans will bring about change in their agricultural systems before environmental degradation forces them to.

Shifting Cultivation in the Contemporary World and Global Environmental Change

Contemporary Shifting Cultivation Practice

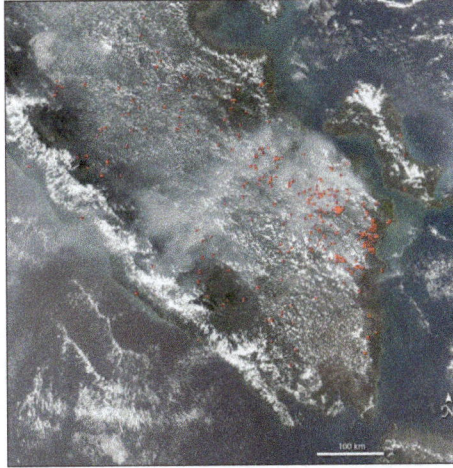

Sumatra, Indonesia

Santa Cruz, Bolivia

The estimated rate of deforestation in Southeast Asia in 1990 was 34,000 km² per year (FAO 1990, quoted in Potter 1993). In Indonesia alone it was estimated 13,100 km² per year were being lost, 3,680 km² per year from Sumatra and 3,770 km² from Kalimantan, of which 1,440 km² were due to the fires of 1982 to 1983. Since those estimates were made huge fires have ravaged Indonesian forests during the 1997 to 1998 El Niño associated drought.

Interdisciplinary Project

Shifting cultivation used to be the backbone of smallholder agriculture throughout the tropics, but today it is abandoned in many places in favor of large scale cash crop production – e.g. for biofuels. The extent of these changes is not well documented because shifting cultivation land rarely appears on official maps and census data seldom identifies shifting cultivators. Moreover, the consequences of these changes for livelihoods (e.g.

food security) are not well known. The aim of this project is to analyze the extent and consequences of change in shifting cultivation by combining meta-analyses of existing studies and census data with case studies in selected areas. This interdisciplinary project focuses on:

1. Trends in change in shifting cultivation landscapes and demography; and

2. Changes in livelihoods due to these changes.

The project will compile data for eight countries (Mexico, Brazil, Laos, Vietnam, Malaysia, Thailand, Zambia and Tanzania) and the outcome is expected to be relevant to planning and policy-making on land and forest management.

Shifting cultivation was assessed by the FAO to be one a causes of deforestation while logging was not. The apparent discrimination against shifting cultivators caused a confrontation between FAO and environmental groups, who saw the FAO supporting commercial logging interests against the rights of indigenous people (Potter 1993, 108). Other independent studies of the problem note that despite lack of government control over forests and the dominance of a political elite in the logging industry, the causes of deforestation are more complex. The loggers have provided paid employment to former subsistence farmers. One of the outcomes of cash incomes has been rapid population growth among indigenous groups of former shifting cultivators that has placed pressure on their traditional long fallow farming systems. Many farmers have taken advantage of the improved road access to urban areas by planting cash crops, such as rubber or pepper as noted above. Increased cash incomes often are spent on chain saws, which have enabled larger areas to be cleared for cultivation. Fallow periods have been reduced and cropping periods extended. Serious poverty elsewhere in the country has brought thousands of land hungry settlers into the cut over forests along the logging roads. The settlers practice what appears to be shifting cultivation but which is in fact a one-cycle slash and burn followed by continuous cropping, with no intention to long fallow. Clearing of trees and the permanent cultivation of fragile soils in a tropical environment with little attempt to replace lost nutrients may cause rapid degradation of the fragile soils.

The loss of forest in Indonesia, Thailand, and the Philippines during the 1990s was preceded by major ecosystem disruptions in Vietnam, Laos and Cambodia in the 1970s and 1980s caused by warfare. Forests were sprayed with defoliants, thousands of rural forest dwelling people uproots from their homes and moved and roads driven into previously isolated areas. The loss of the tropical forests of Southeast Asia is the particular outcome of the general possible outcomes described by Ellen when small local ecological and social systems become part of larger system. When the previous relatively stable ecological relationships are destabilized, degradation can occur rapidly. Similar descriptions of the loss of forest and destruction of fragile ecosystems could be provided from the Amazon Basin, by large scale state sponsored colonization forest land (Becker 1995, 61) or from the Central Africa where what endemic armed conflict is destabilizing rural settlement and farming communities on a massive scale.

Comparison With other Ecological Phenomena

In the tropical developing world, shifting cultivation in its many diverse forms, remains a pervasive practice. Shifting cultivation was one of the very first forms of agriculture practiced by humans and its survival into the modern world suggests that it is a flexible and highly adaptive means of production. However, it is also a grossly misunderstood practice. Many casual observers cannot see past the clearing and burning of standing forest and do not perceive often ecologically stable cycles of cropping and fallowing. Nevertheless, shifting cultivation systems are particularly susceptible to rapid increases in population and to economic and social change in the larger world around them. The blame for the destruction of forest resources is often laid on shifting cultivators. But the forces bringing about the rapid loss of tropical forests at the end of the 20th century are the same forces that led to the destruction of the forests of Europe, urbanization, industrialization, increased affluence, populational growth and geographical expansion and the application the latest technology to extract ever more resources from the environment in pursuit of wealth and political power by competing groups. However we must know that those who practice Agriculture are at the receiving end of the social stratum.

Studies of small, isolated and pre-capitalist groups and their relationships with their environments suggests that the roots of the contemporary problem lie deep in human behavioral patterns, for even in these simple societies, competition and conflict can be identified as the main force driving them into contradiction with their environments.

Alternative Practice in the Pre-Columbian Amazon Basin

Slash-and-char, as opposed to slash-and-burn, may create self-perpetuating soil fertility that supports sedentary agriculture, but the society so sustained may still be over-turned, as above.

Intercropping

Intercropping is a multiple cropping practice involving growing two or more crops in proximity. The most common goal of intercropping is to produce a greater yield on a given piece of land by making use of resources that would otherwise not be utilized by a single crop. Careful planning is required, taking into account the soil, climate, crops, and varieties. It is particularly important not to have crops competing with each other for physical space, nutrients, water, or sunlight. Examples of intercropping strategies are planting a deep-rooted crop with a shallow-rooted crop, or planting a tall crop with a shorter crop that requires partial shade. Inga alley cropping has been proposed as an alternative to the ecological destruction of slash-and-burn farming.

When crops are carefully selected, other agronomic benefits are also achieved. Lodging-prone plants, those that are prone to tip over in wind or heavy rain, may be given structural support by their companion crop. Creepers can also benefit from structural support. Some plants are used to suppress weeds or provide nutrients. Delicate or light-sensitive plants may be given shade or protection, or otherwise wasted space can be utilized. An example is the tropical multi-tier system where coconut occupies the upper tier, banana the middle tier, and pineapple, ginger, or leguminous fodder, medicinal or aromatic plants occupy the lowest tier.

Intercropping of compatible plants also encourages biodiversity, by providing a habitat for a variety of insects and soil organisms that would not be present in a single-crop environment. This in turn can help limit outbreaks of crop pests by increasing predator biodiversity. Additionally, reducing the homogeneity of the crop increases the barriers against biological dispersal of pest organisms through the crop.

The degree of spatial and temporal overlap in the two crops can vary somewhat, but both requirements must be met for a cropping system to be an intercrop. Numerous types of intercropping, all of which vary the temporal and spatial mixture to some degree, have been identified. These are some of the more significant types:

Chili pepper intercropped with coffee in Colombia's southwestern Cauca Department

Coconut and *Tagetes erecta*, a multilayer cropping in India

- Mixed intercropping, as the name implies, is the most basic form in which the component crops are totally mixed in the available space.

- Row cropping involves the component crops arranged in alternate rows. Variations include alley cropping, where crops are grown in between rows of trees, and strip cropping, where multiple rows, or a strip, of one crop are alternated with multiple rows of another crop. A new version of this is to intercrop rows of solar photovoltaic modules with agriculture crops. This practice is called agrivoltaics.

- Temporal intercropping uses the practice of sowing a fast-growing crop with a slow-growing crop, so that the fast-growing crop is harvested before the slow-growing crop starts to mature.

- Further temporal separation is found in relay cropping, where the second crop is sown during the growth, often near the onset of reproductive development or fruiting, of the first crop, so that the first crop is harvested to make room for the full development of the second.

References

- Coleman, Eliot (1995), The New Organic Grower: A Master's Manual of Tools and Techniques for the Home and Market Gardener (2nd ed.), pp. 65, 108, ISBN 978-0930031756.

- Horne, Paul Anthony (2008). Integrated pest management for crops and pastures. CSIRO Publishing. p. 2. ISBN 978-0-643-09257-0.

- Vogt G (2007). Lockeretz W, ed. Chapter 1: The Origins of Organic Farming. Organic Farming: An International History. CABI Publishing. pp. 9–30. ISBN 9780851998336.

- Pamela Ronald; Raoul Admachak (April 2008). "Tomorrow's Table: Organic Farming, Genetics and the Future of Food". Oxford University Press. ISBN 0195301757.

- Henckel, Laura (20 May 2015). "Organic fields sustain weed metacommunity dynamics in farmland landscapes". Proceedings of the Royal Society B. Retrieved 28 February 2016.

- "USDA List of Allowed and Prohibited Substances in Organic Agriculture". USDA List of Allowed and Prohibited Substances in Organic Agriculture. USDA. 4 April 2016. Retrieved 6 April 2016.

- "Can organic food feed the world? New study sheds light on debate over organic vs. conventional agriculture". Science Daily. Retrieved 2 March 2014.

- De Schutter, Olivier. "Report submitted by the Special Rapporteur on the right to food" (PDF). United Nations. Retrieved 3 March 2014.

- Undersander, Dan; et al. "Pastures for Profit: A Guide to Rotational Grazing" (PDF). University of Wisconsin. Cooperative extension publishing. Retrieved 24 February 2014.

- Undersander, Dan; et al. "Grassland Birds: Fostering Habitats Using Rotational Grazing" (PDF). University of Wisconsin. Cooperative extension publishing. Retrieved 24 February 2014.

Biodynamic Agriculture: An Emerging Practice of Agroecology

Biodynamic agriculture is very similar to organic farming. The common features it shares with organic farming are its use of manures and composts. The methods that are unique to biodynamic agriculture are the approaches it takes for the treatment of crops and soil, and its emphasis on local production. This chapter will provide an integrated understanding of biodynamic agriculture.

Biodynamic Agriculture

Biodynamic agriculture is a form of alternative agriculture very similar to organic farming, but which includes various esoteric concepts drawn from the ideas of Rudolf Steiner (1861–1925). Initially developed in the 1920s, it was the first of the organic agriculture movements. It treats soil fertility, plant growth, and livestock care as ecologically interrelated tasks, emphasizing spiritual and mystical perspectives.

Biodynamics has much in common with other organic approaches – it emphasizes the use of manures and composts and excludes the use of artificial chemicals on soil and plants. Methods unique to the biodynamic approach include its treatment of animals, crops, and soil as a single system; an emphasis from its beginnings on local production and distribution systems; its use of traditional and development of new local breeds and varieties; and the use of an astrological sowing and planting calendar. Biodynamic agriculture uses various herbal and mineral additives for compost additives and field sprays; these are sometimes prepared by controversial methods, such as burying ground quartz stuffed into the horn of a cow, which are said to harvest "cosmic forces in the soil", that are more akin to sympathetic magic than agronomy.

As of 2016 biodynamic techniques were used on 161,074 hectares in 60 countries. Germany accounts for 45% of the global total; the remainder average 1750 ha per country. Biodynamic methods of cultivating grapevines have been taken up by several notable vineyards. There are certification agencies for biodynamic products, most of which are members of the international biodynamics standards group Demeter International.

No difference in beneficial outcomes has been scientifically established between certified biodynamic agricultural techniques and similar organic and integrated farming practices. Critics have characterized biodynamic agriculture as pseudoscience on the basis of a lack of strong evidence for its efficacy and skepticism about aspects characterized as magical thinking.

History

Biodynamics was the first modern organic agriculture. Its development began in 1924 with a series of eight lectures on agriculture given by philosopher Rudolf Steiner at Schloss Koberwitz in Silesia, Germany, (now Kobierzyce in Poland southwest of Wrocław). These lectures, the first known presentation of organic agriculture, were held in response to a request by farmers who noticed degraded soil conditions and a deterioration in the health and quality of crops and livestock resulting from the use of chemical fertilizers. The one hundred eleven attendees, less than half of whom were farmers, came from six countries, primarily Germany and Poland. The lectures were published in November 1924; the first English translation appeared in 1928 as *The Agriculture Course.*

Steiner emphasized that the methods he proposed should be tested experimentally. For this purpose, Steiner established a research group, the "Agricultural Experimental Circle of Anthroposophical Farmers and Gardeners of the General Anthroposophical Society". This research group attracted, in the interval 1924 to 1939, about 800 members from around the world, including Europe, the Americas and Australasia. Another group, the "Association for Research in Anthroposophical Agriculture" (Versuchsring anthroposophischer Landwirte), directed by the German agronomist Erhard Bartsch, was formed to test the effects of biodynamic methods on the life and health of soil, plants and animals; the group published a monthly journal *Demeter*. Bartsch was also instrumental in developing a sales organisation for biodynamic products, Demeter, which still exists today. The Research Association was renamed The Imperial Association for Biodynamic Agriculture (Reichsverband für biologisch-dynamische Wirtschaftsweise) in 1933. It was dissolved by the National Socialist regime in 1941. In 1931 the association had 250 members in Germany, 109 in Switzerland, 104 in other European countries and 24 outside Europe. The oldest biodynamic farms are the Wurzerhof in Austria and Marienhöhe in Germany.

In 1938, Ehrenfried Pfeiffer's text *Bio-Dynamic Farming and Gardening* was published in five languages – English, Dutch, Italian, French, and German; this became the standard work in the field for several decades. In July 1939, at the invitation of Walter James, 4th Baron Northbourne, Pfeiffer travelled to the UK and presented the Betteshanger Summer School and Conference on Biodynamic Farming' at Northbourne's farm in Kent. The conference has been described as the 'missing link' between biodynamic agriculture and organic farming because, in the year after Betteshanger, Northbourne published his manifesto of organic farming, *Look to the Land*, in which he coined the term 'organic farming' and praised the methods of Rudolf Steiner. In the 1950s, Hans Mueller was encouraged by Steiner's work to create the organic-biological farming method in Switzerland; this later developed to become the largest certifier of organic products in Europe, *Bioland*.

Today biodynamics is practiced in more than 50 countries worldwide and in a variety of circumstances, ranging from temperate arable farming, viticulture in France, cotton production in Egypt, to silkworm breeding in China. Germany accounts for nearly half of the world's biodynamic agriculture. Demeter International is the primary certification agency for farms and gardens using the methods.

Geographic Developments

- In Australia, the first biodynamic farmer was Ernesto Genoni who in 1928 joined the Experimental Circle of Anthroposophical Farmers and Gardeners, followed soon after by his brother Emilio Genoni. Bob Williams presented the first public lecture in Australia on biodynamic agriculture on 26 June 1938 at the home of the architects Walter Burley Griffin and Marion Mahony Griffin at Castlecrag, Sydney. Since the 1950s research work has continued at the Biodynamic Research Institute (BDRI) in Powelltown, near Melbourne under the direction of Alex Podolinsky. In 1989 Biodynamic Agriculture Australia was established, as a not for profit association.

- In 1928 the *Anthroposophical Agricultural Foundation* was founded in England; this is now called the *Biodynamic Agriculture Association*. In 1939, Britain's first biodynamic agriculture conference, the Betteshanger Summer School and Conference on Biodynamic Agriculture, was held at Lord Northbourne's farm in Kent; Ehrenfried Pfeiffer was the lead presenter.

- In the United States, the Biodynamic Farming & Gardening Association was founded in 1938 as a New York state corporation.

- In France the International Federation of Organic Agriculture Movements (IFOAM) was formed in 1972 with five founding members, one of which was the Swedish Biodynamic Association.

- The University of Kassel had a Department of Biodynamic Agriculture from 2006 to March 2011.

Biodynamic Method of Farming

In common with other forms of organic agriculture, biodynamic agriculture uses management practices that are intended to "restore, maintain and enhance ecological harmony." Central features include crop diversification, the avoidance of chemical soil treatments and off-farm inputs generally, decentralized production and distribution, and the consideration of celestial and terrestrial influences on biological organisms. The Demeter Association recommends that "(a) minimum of ten percent of the total farm acreage be set aside as a biodiversity preserve. That may include but is not limited to forests, wetlands, riparian corridors, and intentionally planted insectaries. Diversity in crop rotation and perennial planting is required: no annual crop can be planted in the same field for more than two years in succession. Bare tillage year round is prohibited so land needs to maintain adequate green cover."

The Demeter Association also recommends that the individual design of the land "by the farmer, as determined by site conditions, is one of the basic tenets of biodynamic agriculture. This principle emphasizes that humans have a responsibility for the development of their ecological and social environment which goes beyond economic aims and the principles of descriptive ecology." Crops, livestock, and farmer, and "the entire socioeconomic environment" form a unique interaction, which biodynamic farming tries to "actively shape ...through a variety of management practices. The prime objective is always to encourage healthy conditions for life": soil fertility, plant and animal health, and product quality."The farmer seeks to enhance and support the forces of nature that lead to healthy crops, and rejects farm management practices that damage the environment, soil plant, animal or human health....the farm is conceived of as an organism, a self-contained entity with its own individuality," holistically conceived and self-sustaining. "Disease and insect control are addressed through botanical species diversity, predator habitat, balanced crop nutrition, and attention to light penetration and airflow. Weed control emphasizes prevention, including timing of planting, mulching, and identifying and avoiding the spread of invasive weed species."

Biodynamic agriculture differs from many forms of organic agriculture in its spiritual, mystical, and astrological orientation. It shares a spiritual focus, as well as its view toward improving humanity, with the "nature farming" movement in Japan. Important features include the use of livestock manures to sustain plant growth (recycling of nutrients), maintenance and improvement of soil quality, and the health and well being of crops and animals. Cover crops, green manures and crop rotations are used extensively and the farms to foster the diversity of plant and animal life, and to enhance the biological cycles and the biological activity of the soil.

Biodynamic farms often have a cultural component and encourage local community, both through developing local sales and through on-farm community building activities. Some biodynamic farms use the Community Supported Agriculture model, which has connections with social threefolding.

Compared to non-organic agriculture, BD farming practices have been found to be more resilient to environmental challenges, to foster a diverse biosphere, and to be more energy efficient, factors Eric Lichtfouse describes being of increasing importance in the face of climate change, energy scarcity and population growth.

Biodynamic Preparations

In his "agricultural course" Steiner prescribed nine different preparations to aid fertilization, and described how these were to be prepared. Steiner believed that these preparations mediated terrestrial and cosmic forces into the soil. The prepared substances are numbered 500 through 508, where the first two are used for preparing fields, and seven are used for making compost. A long term trial (DOK experiment) evaluating the biodynamic farming system in comparison with organic and conventional farming systems, found that both organic farming and biodynamic farming resulted in enhanced soil properties, but had lower yields than conventional farming. Regarding compost development beyond accelerating the initial phase of composting, some positive effects have been noted:

- The field sprays contain substances that stimulate plant growth include cytokinins.

- Some improvement in nutrient content of compost.

Although the preparations have direct nutrient values, their purpose in biodynamics is to support the self-regulating capacities of the soil biota in the case of 500 and 501 and the biological life resident in the composting organics, as well as the mature compost itself, in the others.

Field Preparations

Field preparations, for stimulating humus formation:

- 500: (horn-manure) a humus mixture prepared by filling the horn of a cow with cow manure and burying it in the ground (40–60 cm below the surface) in the autumn. It is left to decompose during the winter and recovered for use the following spring.

- 501: Crushed powdered quartz prepared by stuffing it into a horn of a cow and buried into the ground in spring and taken out in autumn. It can be mixed with 500 but usually prepared on its own (mixture of 1 tablespoon of quartz powder to 250 liters of water) The mixture is sprayed under very low pressure over the crop during the wet season, in an attempt to prevent fungal diseases. It should be sprayed on an overcast day or early in the morning to prevent burning of the leaves.

The application rate of the biodynamic field spray preparations (i.e., 500 and 501) are 300 grams per hectare of horn manure and 5 grams per hectare of horn silica. These are made by stirring the ingredients into 20-50 liters of water per hectare for an hour, using a prescribed method.

Compost Preparations

Compost preparations, used for preparing compost, employ herbs which are frequently used in medicinal remedies. Many of the same herbs are used in organic practices to make foliar fertilizers, turned into the soil as green manure, or in composting. The preparations include:

- 502: Yarrow blossoms (*Achillea millefolium*) are stuffed into urinary bladders from Red Deer (*Cervus elaphus*), placed in the sun during summer, buried in earth during winter and retrieved in the spring.

- 503: Chamomile blossoms (*Matricaria recutita*) are stuffed into small intestines from cattle buried in humus-rich earth in the autumn and retrieved in the spring.

- 504: Stinging nettle (*Urtica dioica*) plants in full bloom are stuffed together underground surrounded on all sides by peat for a year.

- 505: Oak bark (*Quercus robur*) is chopped in small pieces, placed inside the skull of a domesticated animal, surrounded by peat and buried in earth in a place where lots of rain water runs past.

- 506: Dandelion flowers (*Taraxacum officinale*) are stuffed into the mesentery of a cow and buried in earth during winter and retrieved in the spring.

- 507: Valerian flowers (*Valeriana officinalis*) are extracted into water.

- 508: Horsetail (*Equisetum*).

The compost preparations are applied with quantities of 1–2 cm^3 each per 10 m^3 compost, farmyard manure or liquid manure. The preparations should then be evenly sprayed out on the land as soon as possible after stirring.

One to three grams (a teaspoon) of each preparation is added to a dung heap by digging 50 cm deep holes with a distance of 2 meters from each other, except for the 507 preparation, which is stirred into 5 liters of water and sprayed over the entire compost surface. All preparations are thus used in homeopathic quantities. Each compost preparation is designed to guide a particular decomposition process in the composting mass. One study found that the oak bark preparation improved disease resistance in zucchini.

Planting Calendar

The approach considers that there are lunar and astrological influences on soil and plant development—for example, choosing to plant, cultivate or harvest various crops based on both the phase of the moon and the zodiacal constellation the moon is passing through, and also depending on whether the crop is the root, leaf, flower, or fruit of the plant. This aspect of biodynamics has been termed "astrological" in nature.

Seed Production

Biodynamic agriculture has focused on the open pollination of seeds (with farmers thereby generally growing their own seed) and the development of locally adapted varieties. The seed stock is not controlled by large, multinational seed companies.

Biodynamic Certification

The biodynamic certification Demeter, created in 1924, was the first certification and labelling system for organic production. To receive certification as a biodynamic farm, the farm must meet the following standards: agronomic guidelines, greenhouse management, structural components, livestock guidelines, and post harvest handling and processing procedures.

The term *Biodynamic* is a trademark held by the Demeter association of biodynamic farmers for the purpose of maintaining production standards used both in farming and processing foodstuffs.(This is not a trademark held privately in New Zealand) The trademark is intended to protect both the consumer and the producers of biodynamic produce. Demeter International an organization of member countries; each country has its own Demeter organization which is required to meet international production standards (but can also exceed them). The original Demeter organization was founded in 1928; the U.S. Demeter Association was formed in the 1980s and certified its first farm in 1982. In France, Biodivin certifies biodynamic wine. In Egypt, SEKEM has created the Egyptian Biodynamic Association (EBDA), an association that provides training for farmers to become certified. As of 2006, more than 200 wineries worldwide were certified as biodynamic; numerous other wineries employ biodynamic methods to a greater or lesser extent.

Effectiveness

Research into biodynamic farming has been complicated by the difficulty of isolating the distinctively biodynamic aspects when conducting comparative trials. Consequently, there is no strong body of material that provides evidence of any specific effect.

Since biodynamic farming is a form of organic farming, it can be generally assumed to share its characteristics, including "less stressed soils and thus diverse and highly interrelated soil communities".

A 2009/2011 review found that biodynamically cultivated fields:

- had lower absolute yields than conventional farms, but achieved better efficiency of production relative to the amount of energy used;

- had greater earthworm populations and biomass than conventional farms.

Both factors were similar to the result in organically cultivated fields.

Reception

In a 2002 newspaper editorial, Peter Treue, agricultural researcher at the University of Kiel, characterized biodynamics as pseudoscience and argued that similar or equal results can be obtained using standard organic farming principles. He wrote that some biodynamic preparations more resemble alchemy or magic akin to geomancy.

In a 1994 analysis, Holger Kirchmann, a soil researcher with the Swedish University of Agricultural Sciences, concluded that Steiner's instructions were occult and dogmatic, and cannot contribute to the development of alternative or sustainable agriculture. According to Kirchmann, many of Steiner's statements are not provable because scientifically clear hypotheses cannot be made from his descriptions. Kirchmann asserted that when methods of biodynamic agriculture were tested scientifically, the results were unconvincing. Further, in a 2004 overview of biodynamic agriculture, Linda Chalker-Scott, a researcher at Washington State University, characterized biodynamics as pseudoscience, writing that Steiner did not use scientific methods to formulate his theory of biodynamics, and that the later addition of valid organic farming techniques has "muddled the discussion" of Steiner's original idea. Based on the scant scientific testing of biodynamics, Chalker-Scott concluded "no evidence exists" that homeopathic preparations improve the soil.

In Michael Shermer's *The Skeptic Encyclopedia of Pseudoscience*, Dan Dugan says that the way biodynamic preparations are supposed to be implemented are formulated solely on the basis of Steiner's "own insight". Skeptic Brian Dunning writes "the best way to think of 'biodynamic agriculture' would be as a magic spell cast over an entire farm. Biodynamics sees an entire farm as a single organism, with something that they call a life force."

Florian Leiber, Nikolai Fuchs and Hartmut Spieß, researchers at the Goetheanum, have defended the principles of biodynamics and suggested that critiques of biodynamic agriculture which deny it scientific credibility are "not in keeping with the facts...as they take no notice of large areas of biodynamic management and research." Biodynamic farmers are "charged with developing a continuous dialogue between biodynamic science and the natural sciences *sensu stricto*," despite important differences in paradigms, world views, and value systems.

Philosopher of science Michael Ruse has written that followers of biodynamic agriculture rather enjoy the scientific marginalisation that comes from its pseudoscientific basis, revelling both in its esoteric aspects and the impression that they were in the vanguard of the wider anti-science sentiment that has grown in opposition to modern methods such as genetic modification.

Biointensive Agriculture

Biointensive agriculture is an organic agricultural system that focuses on achieving maximum yields from a minimum area of land, while simultaneously increasing biodiversity and sustaining the fertility of the soil. The goal of the method is long term sustainability on a closed system basis. It is particularly effective for backyard gardeners and smallholder farmers in developing countries, and also has been used successfully on small-scale commercial farms.

History

Many of the techniques that contribute to the biointensive method were present in the agriculture of the ancient Chinese, Greeks, Mayans, and of the Early Modern period in Europe, as well as in West Africa (Tapades of Fouta Djallon) from at least the late 18th century. Alan Chadwick brought together the biodynamic and French intensive gardening methods, as well as his own unique approach, to form what he called the Biodynamic-French Intensive method.

The method was further developed by John Jeavons and Ecology Action into a sustainable 8-step food-raising method officially known as "GROW BIOINTENSIVE®️ Sustainable Mini-Farming". The method now enjoys widespread practice and further development, and according to Ecology Action, has been used in over 140 countries around the world, in almost every climate and soil where food is grown. Components important to the biointensive approach include:

- Double-Dug, Raised Beds
- Composting
- biointensive Planting
- Companion Planting
- Carbon Farming
- Calorie Farming
- Use of Open-Pollinated Seeds
- A Whole-System Farming Method

System

The biointensive method provides many benefits as compared with conventional farming and gardening methods, and is an inexpensive, easily implemented sustainable production method that can be used by people who lack the resources (or desire) to implement commercial chemical and fossil-fuel-based forms of agriculture.

Ecology Action's research (Jeavons, J.C., 2001. *Biointensive Mini-Farming* Journal of Sustainable Agriculture (Vol. 19 (2), 2001, p. 81-83) shows that biointensive methods can enable small-scale farms and farmers to significantly increase food production and income, utilize predominantly local, renewable resources and decrease expense and energy inputs while building fertile topsoil at a rate 60 times faster than in nature (*Worldwide Loss of Soil – and a Possible Solution* Ecology Action, 1996).

According to Jeavons and other proponents, when properly implemented, farmers using biointensive techniques have the potential to:

- Use 67% to 88% less water than conventional agricultural methods.

- Use 50% to 100% less purchased (organic, locally available) fertilizer.

- Use up to 99% less energy than commercial agriculture, while using a fraction of the resources.

- Produce 2 to 6 times more food at intermediate yields, assuming a reasonable level of farmer skill and soil fertility (which increase over time as the method is practiced)

- Produce a 100% increase in soil fertility.

- Reduce by 50% or more the amount of land required to grow a comparable amount of food. This allows more land to remain in a wild state, preserving ecosystem services and promoting genetic diversity.

In order to achieve these benefits, the biointensive method uses an eight-part integrated system of deep soil cultivation ("double-digging") to create raised, aerated beds; intensive planting; companion planting; composting; the use of open-pollinated seeds; and a carefully balanced planting ratio of 60% Carbon-Rich Crops (for compost production) 30% Calorie-Rich Crops (for food) and an optional 10% planted in Income Crops (for sale).

The following outline of the methods approximates the descriptions found in the popular biointensive handbook, *How to Grow More Vegetables (and fruits, nuts, berries, grains and other crops) Than You Ever Thought Possible on Less Land Than You Can Imagine*, by John Jeavons, now in its eighth edition, and in seven languages, including braille.

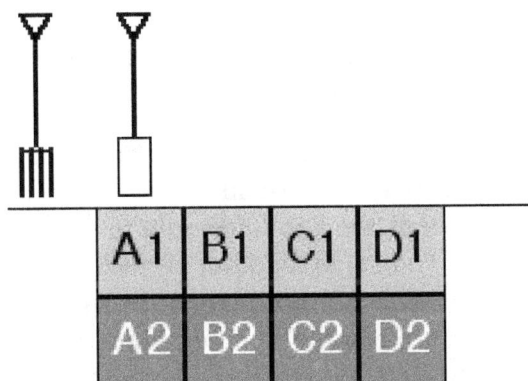

- In double digging, a 12-inch (305 mm) deep trench is dug across the width of the bed with a flat spade, and the soil from that first trench is set aside. The 12 inches (305 mm) below the trench are loosened with a spading fork. When the next trench is dug, that soil is dropped into the empty space of the first trench, and the lower layer is again loosened with a spading fork. This process is repeated along the full length of the bed. The final trench is filled with the soil that was removed from the first trench. The result is a bed that has been tilled to a depth of 24 inches (610 mm). When an entire bed has been double dug, the soil will have greater drainage and aeration, which allows the roots to grow much deeper and reach more nutrients. Despite the fact that no soil has been added, the bed is raised due to the aeration. It is worth noting that hard, unworked soil should be double dug each season until the soil has attained good structure and long lasting aeration. During subsequent seasons, it can be surface cultivated 2 to 4 inches (5 to 10 cm) deep with a hula hoe until compaction again becomes apparent. After double digging the first season, deep tilling during subsequent seasons can be quickly accomplished with a u-bar, particularly in the cases of larger mini-farms or commercial farms.

- Composting allows the plants to transform and enrich the soil with organic matter, and also to return nutrients to the soil. Biointensive composting is fairly straightforward, emphasizing the health and diversity of the microbes that break down and become a part of the compost. Thus, relatively cooler composting is practiced, and plant materials are preferred over animal materials. Soil is often combined with the compost to inoculate the pile with microbes. Without human waste recycling, however, nutrients and organic matter are constantly removed from the soil (as food that is consumed by the farmer) and flushed away. Therefore, when safe and legal human waste recycling is possible—as in many places it already is—that fertility can, and should, be returned to the soil. Another great unappreciated source of compost and soil improvement is the roots of crops themselves, which, in the biointensive system are left to decompose in the soil, where they help to both fertilize and "sew it together", creating stable soil structure. Thus, crops such as alfalfa, which has exceptionally deep roots, and cereal rye, which has a particularly high volume of roots, are valued.

- The soil air from the development of deep soil structure, combined with the microbe- and nutrient-rich compost allow the crops to be planted intensively. To plant intensively, beds are 4 to 6 feet (1.2 to 1.8 m) wide, usually 5 ft (1.5 m) and at least 5 feet (1.5 m) long, often 20 feet (6 m), forming a bed of 100 square feet (10 m²). Crops are not planted in traditional rows according to a square pattern, but are planted in a hexagonal or triangular pattern in the bed so that no space is left unnecessarily unused. These wide beds and close spacings not only allow more plants per area (up to 4 times as many), but also enable the plants to form a living mulch over the soil, keeping in moisture and shading out weeds. Additionally, whenever possible seedlings are started in flats or nursery beds, so that more garden space is available to large plants and so that the seedlings can be more closely spaced before transplant, forming a living mulch in the flat as well.

- Companion planting is described as taking place both in space, which is traditionally called companion planting, and in time, which is traditionally called crop rotation. Companion planting can be used to improve the health and growth of crops, and also as another form of intensive planting, which uses vertical space more efficiently by mixing shallow rooting plants with deep rooting plants or slow growing plants with fast growing plants.

- In order to achieve sustainable fertility on a closed system basis, the biointensive method uses carbon and calorie farming, an aikido-style of work (using the least amount of energy or effort to achieve the greatest amount of work or production), composting—including safe and legal human waste recycling—the use of open pollinated seeds, and limited land use, which allows farmers and gardeners to retain more of the land in a wild state for genetic diversity and an ecosystem balance.

- If carbon or compost crops are grown in about sixty percent of the cultivated land, they can provide the compost materials that maintain the fertility for one hundred percent of the cultivated land. Many cereal crops qualify as compost crops, but provide both food and abundant compost. Some of the compost crops may be grown during the winter, when the land would be otherwise unused. Certain compost crops are higher in carbon while others are higher in nitrogen and/or fix nitrogen in the soil, and the desired proportion of each must be grown for the compost to achieve maximum effectiveness. Also, certain compost crops take particular desired nutrients from the subsoil and concentrate them in the compost, thus allowing a redistribution of those nutrients to the food crops. This proportion of 60% compost crops is crucial to the sustainability that is the goal of the biointensive method, and to the fertility of the garden.

- In calorie farming, care is given to growing enough food energy (and other nutrients) to live on in a minimal area. Root crops are often used in calorie farming because they allow biointensive farmers and gardeners to grow more nutrients in smaller areas, resulting in less labor per calorie, and more space for wilderness and other people. These crops—which have both a high calorie content per pound, and a high yield per area—include potatoes, sweet potatoes, garlic, leeks, burdock, Jerusalem artichoke and parsnips. These crops can produce as much as 5 to 20 times the calories per unit of area per unit of time. In biointensive farming, 30% of the land cultivated for food is used for root crops.

- The use of open pollinated seeds ensures genetic diversity, and allows the farmer to be self-sufficient, harvesting seeds from his or her own plants, and cultivating varieties which are best suited to that particular region.

- The Whole System: biointensive experts emphasize that because these techniques can result in intense productivity and high yields, the system must be practiced as a whole in order to prevent rapid soil exhaustion. The goal of the biointensive method is sustainability, but if the techniques concerning productivity are practiced without integrating the techniques concerning sustainable soil fertility, the soil may be depleted even more rapidly than with conventional farming methods. The most important element for building and maintaining sustainable soil fertility is the growing of 60% compost crops, proper composting techniques that incorporate the right balance of mature carbonaceous brown and green nitrogenous compost materials, and when possible, safe and legal human waste recycling.

Animals

The biointensive method typically concentrates on the vegan diet. This does not mean that biointensive farming must exclude the raising of animals. Animals, while not considered by biointensive practitioners to be sustainable, can be incorporated into biointensive systems, although they increase the amount of land and labor required considerably. The following is excerpted from an article on the topic of integrating animals into a biointensive system from the "Frequently Asked Questions" page on Ecology Action's website:

Livestock can fit into a [biointensive] system, but it usually takes a larger area [than growing a vegan diet]. Normally it takes about 40,000 sq ft of grazing land for 1 cow/steer (for milk/meat) or 2 goats (for milk/meat/wool), or 2 sheep (for milk/meat/wool). [In contrast] With [biointensive farming] and maximizing the edible calorie output in your vegan diet design, one person's complete balanced diet can be grown on about 4,000 sq ft—a much smaller area.

The challenge [to growing animals for food] is that by 2014, 90% of the world's people will only have access to about 4,500 sq ft of farmable land per person, if they leave an equal area in a wild state to protect plant and animal genetic diversity and the world's ecosystems! As you will see from the information that follows on the land requirements for incorporating livestock, this becomes a challenge.

The article goes on to estimate the square footage required to grow fodder for various animals (and compost to replenish the soil), and provides a discussion on whether animal manure should be used as a fertilizer/compost supplement.

Research

Independent research has corroborated Ecology Action's claims that the biointensive system they developed can be sustainable and prolific. Examples include:

- Dr. Ed Glenn at the Environmental Research Lab, University of Arizona, studied the biointensive method when considering food production methods for the Biosphere II experiment. Glenn states that

I had an accidental opportunity to test the [biointensive] method of zero-input agriculture as a part of our work with Biosphere II. [We tested the method] and we published it in Hortscience. Then the second group of Biospherians used those same methods, and they put a publication in Ecological Engineering. So the John Jeavons [biointensive] method not only works,it actually has the scientific stamp of approval. (*Sustainable Food Production for a Complete Diet*, E. Glenn and C. Clement, Environmental Research Laboratory, University of Arizona, Tucson; P. Brannon, Dept. of Food science and Nutrition, U. Arizona, Tucson; L. Leigh, *Space Biospheres Venture*, Oracle, AZ, Hortscience, vol. 25 (12), December, 1990.)

- A 2010 study published in the journal Renewable Agriculture and Food Systems showed that biointensive methods resulted in significantly increased production and a reduction of energy use when compared with conventional agriculture (Moore, S.R., 2010, *Energy efficiency in small-scale biointensive organic onion production in Pennsylvania*, USA, Renewable Agriculture and Food Systems, 25:3, pp. 181-188). This study states that "Current mechanized agriculture has an energy efficiency ratio of 0.9 ... energy efficiency for biointensive production of onions in our study was over 50 times higher than this value (51.5), and 83% of the total energy required is renewable energy."

- In 2010, the UNCCD (United Nations Convention to Combat Desertification) posted an article detailing the benefits of biointensive agriculture, *Grow Biointensive System, a tool to fight against desertification*.

Regenerative Agriculture

Biodiversity

Regenerative agriculture is a sub-sector practice of organic farming designed to build soil health or to regenerate unhealthy soils. The practices associated with regenerative agriculture are those identified with other approaches to organic farming, including maintaining a high percentage of organic matter in soils, minimum tillage, biodiversity, composting, mulching, crop rotation, cover crops, and green manures.

Regenerative v. Organic Agriculture

Hoverfly at work

In the past, regenerative farming was seen as a long-term integrated approach that proponents used to build soil health, promote nutrient retention, and encourage pest and disease resistance. Many of the practices associated with regenerative farming are management practices associated with organic agriculture . In practice, these practices can be applied in any type of horticulture and properly managed livestock with Holistic Planned Grazing (Savory & Butterfield Holistic Management -A new decision making framework) where one of the main goals is to build soil organic matter, an organic practice understood by practitioners of organic farming to have far reaching benefits for plant health and farm sustainability. When combined with the spirit of organic agriculture such practices are said to produce healthy soil, healthy food, clean water and clean air using inexpensive inputs local to the farm. Practices that minimize biota disturbance and erosion losses while incorporating carbon rich amendments and retaining the biomass of roots and shoots are encouraged in regenerative farming.

Best Practices

Foremost among best practices in regenerative farming are zero-tolerance for synthetic pesticides, fertilizers, and other inputs that disrupt soil life. On the other hand, conservation tillage, while not yet widely used in organic systems, is viewed as a regenerative organic practice integral to soil-carbon sequestration.

Contributions from the Rodale Institute

Rodale Institute, Test Garden

Recent scientific research has shown that practices inherent in the regenerative philosophy contribute to carbon sequestration by natural processes of photosynthetic removal and retention of atmospheric CO_2 in soil organic matter. At the forefront of this research is the Rodale Institute, which is one of the major proponents of regenerative agriculture. Notably, the concept of *regenerative organic agriculture* was coined by Robert Rodale prior to his untimely death in 1990. The Rodale approach defines regenerative farming as a long-term, holistic design that attempts to grow as much food using as few resources as possible in a way that revitalizes the soil rather than depleting it, while offering a solution to carbon sequestration. The mantra is the slogan: "Healthy Soil = Healthy Food = Healthy People." In Rodale's view, when coupled with the management goal of carbon sequestration, regenerative "farming becomes, once again, a knowledge intensive enterprise, rather than a chemical and capital-intensive one."

Permaculture and Regeneration Science

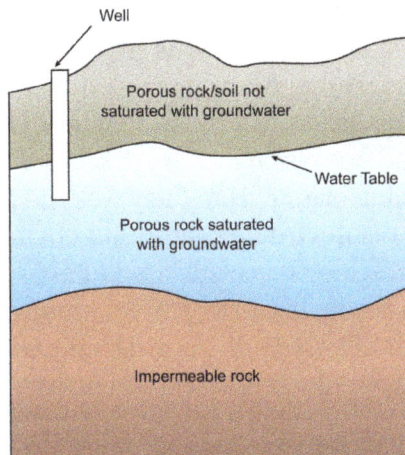

Groundwater

In permaculture, a regenerative farm is one where biological production and ecological structure are growing increasingly more complex over time, but yields continue to increase while external inputs decrease. The way that regeneration is determined in this construct is by whether the components in the system or actions taken in the system increase both biological diversity *and* biomass. The overall health of the ecological system in which the farm and its humans are guests is determined by the health of the water and soil. This is achieved by strengthening (making more resilient, redundant) three key components of regenerative ecosystem management:

1. Increased biomass

2. Increased biological activity

3. Intentional remineralization

Foundations of Regenerative Agriculture

The foundation of regenerative agriculture has occurred over thousands of years, a span of time when most farming was organically practiced worldwide. Even then, its knowledge base was not global because its practitioners excluded traditional non-western local knowledge systems, about which we are only beginning to learn more. It is not possible to list the many contributors or contributions to this process over many centuries. Yet, some voices and systems should be illuminated. Among these, in alphabetical order, are the following:

William Albrecht (1888–1974), an agronomist at the University of Missouri for many years. Through his writings, lectures, and radio programs, Albrecht promoted the intimate relationship between healthy soil and animal nutrition, a relationship that includes humans. Feed the soil to feed the plants to feed the consumer was his mantra.

Lady Evelyn Barbara "Eve" Balfour (1899–1990) was a founding figure in the 20th century organic movement and an organic farming pioneer. She was one of the first women to study agriculture at an English university, graduating from the University of Reading, and began farming in 1920. By 1939, she had launched a privately funded experimental farm, called Haughley, to test the principles of organic farming. The initial findings of the work there and her research were published in Living Soil (1948), which has become an organic farming classic. Haughley was the first long-term comparative research project measuring results from organic and chemically based farming. What the results of her work at Haughley revealed should have led to a global push to institute sustainable agriculture throughout the British empire and beyond, but that was not to be, as the global power of post-WWII capitalism undermined such efforts.

George Washington Carver (1864–1943), inventor and scientist of Tuskegee University, was the grand-daddy of sustainable agriculture in the USA. His contributions are too many to detail here, but among them is his development of the U.S. agriculture extension system and his many inventions in polymer science. His influence on U.S. agriculture in the first half of the 20th century is not fully appreciated nor, more tragically, even known by contemporary proponents of today's biointensive farming methods. Carver's scientific research was the foundation of US agriculture policy under two generations of Secretaries of Agriculture. His death in 1948 marked the end of organic agricultural practices in commercial agriculture in the United States until recently, as petrochemical companies and Agribusiness corporations competed successfully to control the actors in the various industries.

Alan Chadwick (1909–1980) was a leading innovator of organic farming techniques and an influential educator in the field of biodynamic/French intensive gardening through much of the 20th century. A student of Rudolf Steiner, Chadwick is often cited as inspirational to the development of the "California Cuisine" movement.

George Chan (born 1923), former Ministry of Agro-Industry & Fisheries of Mauritius, and Mae-Wan Ho, Director of the Institute of Science in Society, are architects of Dream Farm, a model of a sustainable zero-emission, zero-waste production farm that maximizes the use of renewable energies, turns residues into food and energy resources, and eliminates the need for fossil fuels. A core principle of Dream Farm is that sustainable systems are organisms, comprising the farmer, livestock and crops. In a Dream Farm, very little input is wasted or exported to the environment; most are recycled and kept inside the system by consciously integrating food and energy production. Mr. Chan has been a pioneer in the field of sustainable recycling of waste using agro-ecological systems. He developed the Integrated Farming and Waste Management System, which involves a sustainable cycle in which matter and energy circulate through different stages, dramatically increasing yields.

Field Hamois Belgium Luc Viatour

Masanobu Fukuoka (1913–2008) of Japan lived a long and dutiful life in partnership with his environment. He was a farmer, activist, and author of the practices and theory of natural farming, on which he based his four core uncompromising principles: no cultivation, no (chemical) fertilizers, no weeding, and no pesticides. Among his teachings is that for a farmer to be successful, he must form a partnership with the natural environment, derive an intimate understanding of it together with the plants a farmer chooses to grow. Features of his philosophy are present in most contemporary farming practices. His "seed balls" cultivation innovation is widely used in many horticultural environments and in commercial retail products, including lawn seed.

Takao Furuno (born 1950) of Japan is the architect of the Aigamo Method, a modernization of an 800-year-old Chinese technique of using ducks to promote sustainable rice cultivation. Funuro's system is polycultural, combining rice cultivation with duck husbandry, aquaculture, and vegetable production. Together, these enterprises provide income from rice, vegetable and flower production, eggs, meat, and live ducks from the aigamo, and fish from the paddies. The aigamo ducks are a breed derived from wild and domestic ducks, whose ducklings provide the labor for cultivation, pest control, and manure for fertilization of the rice paddies. This beneficial marriage between duck and rice eliminates dependency on chemical fertilizers, pesticides, molluscicides, fossil fuels, and heavy duty equipment as costs, while sustaining a safe environment for farmers to work and simultaneously increasing net production and farm income.

John D. Hamaker (1914–1994) was a mechanical engineer, ecologist and practical visionary who from 1968 to 1994 tried to awaken the world to the crucial need to remineralize the soils and regenerate the Earth's life-support system. His motivations included a profound desire to help create a healthy, just and real civilization rooted in ecological wisdom, and his realization that malnutrition and disease followed by famine and glaciation could be ended by a total human commitment.

Julius Hensel (1844–1903), a German miller and author of Bread from Stones. In the 1890s, Hensel was an early advocate of restoring trace minerals to soil with dust from primeval stones and reported successful results with his steinmehl (stonemeal). It is said that his ideas were not accepted due to both technical limitations and financing. But, according to proponents of his method, Hensel's opposition from manufacturers of chemical fertilizers, set the stage for what would happen eventually in American agriculture following WWII when there was little opposition and the rise of the U.S. petrochemical industry. Today, Hensel's pioneer work in opposing the use of chemicals in agriculture found rebirth in the Organic Movement a half a century later. Yet, Hensel may be more modern than the most modern agricultural reformer. On the basis of theoretical chemical considerations, supported by practical tests, he claimed that his rock dust can replace not only chemical fertilizers, but all animal ones as well.

Sepp Holzer (born 1942) of Austria, and his wife Veronika, have created a diverse and natural way of growing food in an unconventional way by using a terraced system of mounds on the Austrian mountainsides, referred to as hugelkultur. The mounds are built on a foundation of organic materials, a traditional way of growing in the region of The Krameterhof in Lungau, Austria, just not at 1000+ meters above sea level. Holzer's edible microclimates are considered one of the few perfectly working permaculture systems in the world. After almost 40 years of continuous production, the Holzer farms contain a complex of pond culture, terraces, water power station, thousands of fruit trees cultivated among companionable plant families, thirty different types of potatoes, many different grains, fruits, vegetables, herbs and wildflowers are growing just about everywhere — in the forest, on extremely steep hills, on rocky outcrops, on stone pathways, and around ponds, all without the use of any pesticides, herbicides or synthetic fertilizers. Their video, *Farming With Nature: A Case Study of Successful Temperate Permaculture*, is widely distributed on the Internet and a must-view for anyone looking to develop a sustainable collaborative system of food production.

Sir Albert Howard (1873–1947), an English botanist, was an agricultural advisor to the British government in charge of a colonial research farm at Indore in India. Howard has been called the father of modern composting for his refinement of a traditional Indian composting system into what is now known as the Indore method. It was at Indore that Howard documented and tested Indian organic farming techniques. Sir Howard shared this knowledge through the Soil Association in England and the Rodale Institute in the United States. In his later years, he was the editor of the influential journal, *Biodynamics Journal*.

Elaine Ingham, a microbiologist and founder of Soil Foodweb, Inc. in 1996, is recognized as a premier authority in soil microbiology and the soil food web. Through her pioneering research and lectures, Dr. Ingham has been instrumental in popularizing the importance of soil health and the growing public understanding of the soil food web in sustaining this health. Since January 2011, she has been the Chief Scientist at The Rodale Institute where she continues to study the microbial life of the soil and to give lectures on her findings.

John Ikerd, a successor to William Albrecht and professor emeritus at the University of Missouri, continues to be a staunch advocate for the "small" family farm and farmers. Today, he is an active crusader for sustainability in the US food system. His views and writings are available on his Website and at YouTube. Dr. Ikerd is author of *The essentials of economic sustainability*, *Small Farms are Real Farms: Sustaining People through Agriculture* and *Sustainable Capitalism* (2005).

John Jeavons, the inspiration and architect of a sustainable 8-step food production method officially known as Grow Biointensive, which combines elements of French intensive and biodynamics techniques. The Grow Biointensive approach is promoted by Ecology Action, a non-profit that operates a research mini-farm in Willits, CA and a retail store called Bountiful Gardens. Both projects promote the Grow Biointensive method teaching people in more than a hundred countries. Ecology Action's research and publications have several goals: (1) enabling small-scale farms and farmers worldwide to significantly increase food production and income by (2) utilizing predominantly local, renewable resources to decrease expenses and energy inputs (labor, land, water) and (3) building fertile topsoil at a rate 60 times faster than in nature.

Patricia Lanza was born in 1935 in Crossville, Tennessee, to teenagers George and Mamie Neal. An only child for eight years, she spent her early formative years with grandparents while her parents worked in Detroit, MI. She perfected and authored several books about lasagna gardening or sheet mulch gardening, a type of gardening perfected by Ruth Stout in the 1950s, but whose books had gone out of print. Lanza introduced a new generation of gardeners to the ultimate no-till method of growing in which little or no labor is wasted digging and amending soils. Instead, the lasagna method feeds the soil biota from above and encourages the soil food web to do the work of aerating and mixing the nutrients into the soil below.

Jacob Mittleider, architect of the Mittleider Method, a popular contemporary method of soil-less growing. Mittleider's system is being continued and refined by Jim Kennard at the Food for Everyone Foundation.

Bill Mollison and David Holmgren, architects of Permaculture or "permanent agriculture," a holistic approach that combines ecological design with natural principles of horticultural production. Both men are ardent advocates for creating communities that work in harmony with nature rather than in opposition to it. Permaculture is a worldwide phenomenon whose principles and practices have been published in many of the world's languages and is rapidly being integrated into public planning projects across the globe. Video presentations about Permaculture and its practitioners are broadly distributed on the Internet and can be found at YouTube.

Maynard Murray (1910–1983) was a medical doctor and a pioneer in merging the disciplines of biology, health and agriculture from the 1930s when he began experimenting with "sea-solids"–mineral salts that remain after total sea water evaporation. Around 1940, he began to perform extensive experiments to determine when the proportions of trace minerals and other elements present in sea water were optimum for growth and health of both land and sea life. His extensive experiments demonstrated repeatedly and conclusively that plants fertilized with sea solids and animals fed sea-solid-fertilized feeds grow stronger and more resistant to disease. Murray recounts his experiments and presents his conclusions in his classic work, Sea Energy Agriculture (1976; republished in 2001). Largely ignored during his lifetime, his lifelong quest contributed greatly to our understanding of the role of trace minerals in the healthy growth of all organisms on the planet, including humans.

J.I. Rodale (1898–1971) was an early proponent of organic and regenerative farming and founder of the Rodale Institute in the United States. He is credited with launching organic gardening practices more broadly in the United States through his writings, research, and publishing enterprise. The Rodale enterprises continue to make contributions around the world through their advocacy, research, demonstrations, and publishing.

Robert Rodale (1930–1990), former CEO of The Rodale Institute, was a major advocate of regenerative agriculture, fostered the Regenerative Agriculture Association, published numerous books on the subject, funded research, established demonstration fields, sponsored practitioners in the field, and spread the knowledge system of regenerative agriculture around the globe. He coined the concept of 'regenerative organic agriculture' to distinguish it from 'sustainable' agriculture.

Bhaskar H. Save (born 1922) is creator of the highly successful Kalpavruksha ("wish-fulfilling tree") Farm in Umbergaon, India established in 1953. After practicing traditional agriculture for many years with poor results, Sri Save committed his resources to organic farming and developed a system of natural farming that Masanobu Fukuoka, the noted founder of natural farming, praised as the best example of natural farming he had witnessed anywhere. Sri Save used intensive interplanting in which short life-span vegetables (alpa-jeevi), medium life-span species (madhya-jeevi – such as banana, papaya, and custard apple), and long life-span species (deergha-jeevi–such as chikoo, coconut, mango) are combined and phased in over time until the long life-span species mature.

Ethan Roland Soloviev and Gregory Landua, cofounders of Terra Genesis International (a regenerative agriculture and supply company), published a paper in 2016 titled *Levels of Regenerative Agriculture*. In this paper, they describe a four-fold framework consisting of:

- Functional Regenerative Agriculture: "humans can do good through their agricultural production"

- Integrative Regenerative Agriculture: "grow the health and vitality of the whole ecosystem"

- Systemic Regenerative Agriculture: requiring personal development; "farms are woven into an ecosystem of enterprises operating in their bioregion"

- Evolutionary Regenerative Agriculture: requiring pattern understanding; "harmonize with the potential of a place," and "develop a diversity of global and local regenerative producer webs"

Rather than creating a hierarchy, Soloviev and Landua posit that each level of regenerative agriculture has its place, depending upon context and aim.

Rudolf Steiner (1861–1925), an Austrian and one of the great minds of the 20th century, is the architect of the Biodynamics method of agriculture, the first truly modern ecological food production system. Biodynamic agriculture was one of the first systems also to treat farms as unified and individual organisms or ecosystems. The system has its basis in a world-view referred to as Anthroposophy, which Steiner developed from his marriage of spiritual, cultural, and intellectual life experiences. Biodynamics emphasizes a holistic balance between the soil, plants, and animals in a closed, self-nourishing system, with an emphasis on animal waste products and composts and exclusion of artificial chemicals. Some approaches are unique, such as (1) the use of fermented herbal and mineral preparations as compost additives and field sprays and (2) the use of an astronomical sowing and planting calendar. The biodynamics method continues to have a large following worldwide, and deservedly so—Rudolf Steiner was a brilliant thinker in many ways.

Ruth Stout (1884–1980) lived a long, active and productive life. By the 1950s, she had perfected a "no-till" method of gardening that she promoted as "no work" in her writings about gardening, including two books, *How to Have a Green Thumb Without an Aching Back* and *Gardening Without Work for the Aging, the Busy, and the Indolent*. The latter volume was republished by Mother Earth News in 2011. Her work has led to other innovations in no-till practices, such as *slash and mulch* in the tropics.

Charles Walters (1926–2009) was an economist, journalist, farmers advocate in the first phase of his career with the National Farmer's Organization; and founder, publisher and editor of Acres U.S.A., North America's oldest publisher on production-scale organic and sustainable farming in the second phase of his extraordinary life. Walters penned hundreds of articles on the technologies of organic and sustainable agriculture and is author or co-author of several books, including *Eco-Farm, Weeds: Control Without Poisons, Unforgiven*, a book about visionary farm economist Carl Wilken, and many more. In 1970, shortly after he started Acres, Walters coined the term "eco-agriculture" because he wanted to unify the concepts of "ecological" and "economical" in the belief that unless agriculture was ecological, it could not be economical.

Keyline Irrigation, Taranaki Farm

Don Weaver, a protégé of John Hamaker, is an ecologist and gardener, who assisted Hamaker in advocating for policies and practices of soil remineralization, biosphere regeneration, and climatic stabilization. He continues to promote these causes today.

Booker T. Whatley (1915–2005), a horticulturalist and beneficiary of the George Washington Carver tradition, may be best remembered for popularizing U-Pick farms and their direct marketing approach through fee-based customer subscriptions. But, he was also among the first practitioners of sustainable agriculture to focus more directly on the economic concerns of small farmers, encouraging them to identify high value crops and enterprises that were more profitable on smaller units of land, such as shiitake mushrooms, the husbandry of small ruminants, specialty cheeses, and much more.

P.A. Yeomans (1904–1984), an Australian geologist and the architect of the Keyline design, an innovative solution to farm water management, is little known outside of Australia for his many contributions from sustainable agriculture to soil fertility to farm management in the early 20th century. His use of land topography to harvest rain water into ponds is quietly used today in the construction of swales and berms in production units ranging from backyard gardens to the monumental landscaping practices of Sepp Holzer today.

References

- Lejano RP, Ingram M, Ingram HM (2013). "Chapter 6: Narratives of Nature and Science in Alternative Farming Networks". Power of Narrative in Environmental Networks. MIT Press. p. 155. ISBN 9780262519571.

- Ikerd, John (2010). "Sustainability, Rural". In Leslie A. Duram. Encyclopedia of Organic, Sustainable, and Local Food. ABC-CLIO. pp. 347–349. ISBN 0313359636.

- Abbott, L. K.; Murphy, Daniel V. (2007). Soil Biological Fertility: A Key to Sustainable Land Use in Agriculture. Springer. p. 233. ISBN 140206618X.

- Goode, Jamie (2006-03-01). The science of wine: from vine to glass. University of California Press. ISBN 978-0-520-24800-7.

- K. Padmavathy; G. Poyyamoli (2011). Lichtfouse, Eric, ed. Genetics, biofuels and local farming system. Berlin: Springer. p. 387. ISBN 978-94-007-1520-2.

- Desai, B K (2007). Sustainable agriculture: a vision for the future. New Delhi: B T Pujari/New India Pub. Agency. pp. 228–9. ISBN 978-81-89422-63-9.

- Dugan D (2002). Shermer M, ed. Anthroposophy and Anthroposophical Medicine. The Skeptic Encyclopedia of Pseudoscience. 2. ABC-CLIO. p. 32. ISBN 1-57607-653-9.

- John Jeavons, How to Grow More Vegetables: And Fruits, Nuts, Berries, Grains, and Other Crops Than You Ever Thought Possible on Less Land Than You Can Imagine ISBN 1-58008-233-5.

- Vogt G (2007). Lockeretz W, ed. Chapter 1: The Origins of Organic Farming. Organic Farming: An International History. CABI Publishing. pp. 9–30. ISBN 9780851998336.

Permissions

Index

www.ingramcontent.com/pod-product-compliance
Lightning Source LLC
Chambersburg PA
CBHW061934190326
41458CB00009B/2735